WHY DNA?

Information is central to the evolution of biological complexity, a physical system relying on a continuous supply of energy. Biology provides superb examples of the consequent Darwinian selection of mechanisms for efficient energy utilisation. Genetic information, underpinned by the Watson–Crick base-pairing rules, is largely encoded by DNA, a molecule uniquely adapted to its roles in information storage and utilisation. This volume addresses two fundamental questions: First, what properties of the molecule have enabled it to become the predominant genetic material in the biological world today; and second, to what extent have the informational properties of the molecule contributed to the expansion of biological diversity and the stability of ecosystems. The author argues that bringing these two seemingly unrelated topics together enables Schrödinger's *What Is Life?*, published before the structure of DNA was known, to be revisited and his ideas examined in the context of our current biological understanding.

∽

Andrew Travers is an emeritus scientist at the Medical Research Council Laboratory of Molecular Biology (MRC LMB) and a visiting scientist in the Department of Biochemistry at the University of Cambridge. His research focuses on the use of the genetics and biochemistry of bacteria and *Drosophila* to study the mechanisms of chromatin folding and unfolding. He started his academic career at the MRC LMB before spending two years as a postdoc in Jim Watson's lab at Harvard University, where he co-discovered the first of the RNA polymerase sigma factors.

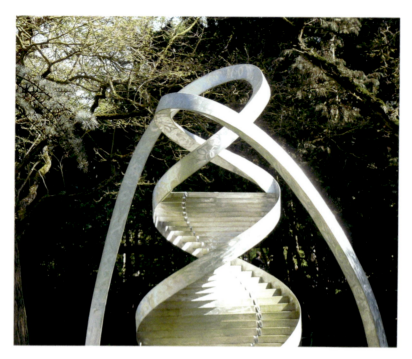

Frontispiece: DNA sculpture. Photograph by Sarah Rodger, reproduced here with permission from Charles Jencks

WHY DNA?

FROM DNA SEQUENCE TO
BIOLOGICAL COMPLEXITY

Andrew Travers

MRC Laboratory of Molecular Biology and University of Cambridge

CAMBRIDGE
UNIVERSITY PRESS

CAMBRIDGE
UNIVERSITY PRESS

University Printing House, Cambridge CB2 8BS, United Kingdom

One Liberty Plaza, 20th Floor, New York, NY 10006, USA

477 Williamstown Road, Port Melbourne, VIC 3207, Australia

314–321, 3rd Floor, Plot 3, Splendor Forum, Jasola District Centre, New Delhi – 110025, India

103 Penang Road, #05–06/07, Visioncrest Commercial, Singapore 238467

Cambridge University Press is part of the University of Cambridge.

It furthers the University's mission by disseminating knowledge in the pursuit of education, learning, and research at the highest international levels of excellence.

www.cambridge.org
Information on this title: www.cambridge.org/9781107056398
DOI: 10.1017/9781107297241

First published 2022

Printed in the United Kingdom by TJ Books Limited, Padstow Cornwall

A catalogue record for this publication is available from the British Library.

Library of Congress Cataloging-in-Publication Data
NAMES: Travers, A. A. (Andrew Arthur), author.
TITLE: Why DNA? : from DNA sequence to biological complexity / Andrew Travers.
DESCRIPTION: Cambridge, United Kingdom ; New York, NY : Cambridge University Press, 2021. | Includes bibliographical references and index.
IDENTIFIERS: LCCN 2020055277 (print) | LCCN 2020055278 (ebook) | ISBN 9781107056398 (hardback) | ISBN 9781107697522 (paperback) | ISBN 9781107297241 (epub)
SUBJECTS: MESH: Schrödinger, Erwin, 1887-1961. What is life? | DNA | Biological Evolution | Biodiversity | Biophysical Phenomena–genetics | Origin of Life
CLASSIFICATION: LCC QP624 (print) | LCC QP624 (ebook) | NLM QU 58.5 | DDC 572.8/6–dc23
LC record available at https://lccn.loc.gov/2020055277
LC ebook record available at https://lccn.loc.gov/2020055278

ISBN 978-1-107-05639-8 Hardback
ISBN 978-1-107-69752-2 Paperback

Contents

Preface

This book was first conceived many years ago over a dinner on the banks of the River Weser with my long-term friend and colleague Georgi Muskhelishvili simply as an exploration of the unique properties of the DNA polymer that could explain its predominance as the currently preferred genetic material. But during the subsequent, very extended, gestation of the concepts, the implications of the capacity of DNA to store information in different ways became more and more apparent. The project morphed in the writing and became a personal intellectual journey to explore the role of DNA as an information carrier in the evolution of complex systems. DNA is precisely that – a carrier or vehicle for two modes of information, and it is that information per se that drives the burgeoning increase in biological complexity.

But first a disclaimer. A book on DNA could cover a vast range of highly topical aspects of modern medical practice, forensics and genetic engineering to name but a few. If you are looking for enlightenment on those and related topics this book is not for you. It is concerned, as the subtitle suggests, with the relationship between DNA, information and complexity and traces the evolution of that relationship from the origin of life.

It is a given that biological systems are subject to the laws of physics, but how do the principles that govern the generation of the immense panoply of biological diversity fit with our understanding of the physical world? More than 60 years ago, in the 1940s, three substantial cornerstones of this problem were put in place. In 1946 in *What Is Life?*, the Nobel prize–winner Erwin Schrödinger proposed that biological systems, essentially complex chemical systems, evolved by using energy to minimise entropy – or in his terminology to increase *negentropy*. At about the same time Claude Shannon developed his 'Mathematical Theory of Communication', derived from considerations of the efficiency of the transmission of information in undersea cables, while John von Neumann distinguished between different ways of encoding information. Meanwhile in the laboratory, Oswald Avery and his collaborators realised that a preparation highly enriched in DNA from the bacterium *Pneumococcus* could transform a rough-coated bacterium to a smooth-coated one, an early example of using DNA for genetic manipulation. In the biological context these strands coalesced with the discovery by Watson and Crick of the structure of DNA and their realisation that the sequence of bases in the molecule could constitute a heritable genetic blueprint. Since then the mechanisms by which the information encoded in DNA is utilised and manipulated have been largely elucidated. But even knowing that the DNA molecule is the fundamental carrier of genetic information in the biological world and how that information is used still leaves interesting questions. What is it about this molecule, and not other similar ones, that has driven its selection as the principal genetic material? At the most fundamental level is it that it is simply slightly different, and therefore distinguishable from the closely related RNA? An equally fundamental issue is why biological systems – molecular assemblies, organisms, ecosystems – tend to increase in diversity and complexity over evolutionary time. This phenomenon necessarily parallels an increase in the amount of information carried by the ensemble of associated DNA molecules. But is there a principle driving these expansions that is compatible with Schrödinger's original insights?

If the evolutionary increase in organisation is driven by energy, its primary source, usually light, is processed by a biological system acting as a whole. But an increase in organisation implies a concomitant increase in information, and ultimately this information is encoded in DNA. And so appreciation of the full implications of *What Is Life?* includes not only an understanding of DNA structure and genetics but also, at the other extreme, of how ecosystems and even complex cosmo- logical systems work. In his preface Schrödinger included the apology:

> But the spread, both in width and depth, of the multifarious branches of knowledge during the last hundred odd years has confronted us with a queer dilemma. We feel clearly that we are only now beginning to acquire reliable material for welding together the sum-total of what is known into a whole; but, on the other hand, it has come next to impossible for a single mind fully to command more than a small specialised portion of it. I can see no other escape from this dilemma (lest our true aim is to be lost forever) than that some of us should venture to embark on a synthesis of facts and theories, albeit with second-hand and incomplete knowledge of some of them, and at the risk of making fools of themselves.

To me this caveat seems even more apposite now and the risks even greater. Biological science now encompasses an even greater number of specialisations, each requiring expert knowledge and accompanied by its own esoteric argot that can only be acquired over many years. I have tried to navigate through this minefield but with what success only the reader can judge. Invariably and necessarily so, in science some of the issues I've discussed may be contentious, but if so, I hope that the views argued here will provide a basis for discussion.

Ultimately this is yet another book about the ever fascinating topic of evolution. Perhaps writing this is symptomatic of a common affliction of biologists for, unlike the inhabitants of idyllic Grantchester,

> when they get to feeling old,
> They think of evolution, I'm told.

Acknowledgements

Just as the Cambridge Botanic Garden, where I now work as a guide (and so finally have become more familiar with the joys of botany), can be thought of as a cradle of evolutionary thought, the Laboratory of Molecular Biology (LMB) was a cradle for molecular biology and, of course, DNA. Importantly, during 44 years there, LMB was a melting pot of creativity. Discussions and collaborations with many fellow scientists – including students and postdocs – shaped my outlook – most notably including those with Horace Drew, whose DNA-centric perspective challenged my more biological approach. My scientific debt to LMB is tremendous. I believe that LMB by example taught me to think as a scientist. Over half a century ago, Francis Crick's crystal-clear exposition of the progress of the 'wobble' hypothesis in successive seminars was especially influential.

And also my gratitude to the wider scientific diaspora. The essence of science is – or should be – that when your ideas are challenged, you are forced to think more clearly and so I am extremely grateful to those who, during my scientific career, took issue with, to them, some of my perhaps more speculative notions. My experience is that the outcome of such apparent disagreement is often that both points of view are largely correct but incomplete. Often scientists think they've done the same experiment as someone else but have actually performed a completely different experiment. The consequent synergistic synthesis then adds substantially to an understanding of the problem. There are so many scientific friends who have sustained and inspired me over the years that I cannot acknowledge them all here, but special mentions must go to the chromatin group in the Department of Biochemistry at Cambridge and to those erstwhile colleagues from my time at Harvard who remain in touch.

In the wider world, included amongst those who have led me onto paths I might not otherwise have taken are Ernesto Di Mauro, who first introduced me to DNA topology and the problem of the origin of life; Henri Buc and Malcolm Buckle, who taught me that physics is crucial to the understanding of biology; and Georgi Muskhelishvili, who encouraged me to think holistically and who more than anyone else was responsible both for the genesis of this book and also for the development of many of the ideas discussed in it. And of course there was always the element of chance – being in the right place at the right time. Also latterly the DNA sculptures and cosmic thinking of the late Charles Jencks have been inspirational. To my great regret, I was never able to discuss the cosmos with him.

I owe an immense amount to the editorial staff of Cambridge University Press, especially for their incredible patience. Katrina Halliday, Aleksandra Serocka and Natasha Whelan in the gentlest way both kept me writing and cushioned me from the consequences of life's events. I am also extremely grateful to those, especially Georgi Muskhelishvili and Ernesto Di Mauro, who braved my prose and read selected draft chapters. Their input is enormously appreciated.

During the writing of this book I have been unfailingly encouraged by the extended Travers pack. All the human members, especially Sarah and Elli, gave me the motivation that I so often lacked, and the canine component invariably provided enthusiastic (and occasionally overenthusiastic) companionship. But most of all I am immensely grateful to Carrie, my late wife. It is no exaggeration to write that without her constant encouragement this book would have remained just a distant dream.

1

The Perennial Question

*Available energy is the main object at stake in the struggle for existence
and the evolution of the world.*
—*Ludwig Boltzmann, 1886*

Gain in entropy always means loss of information, and nothing more.
—*Gilbert Lewis, 1930*

The perennial question – what is life? The simple answer is that life, either considered in the totality of all its incredible diversity or even in the context of an individual organism, is a highly complex chemical system with a capacity for self-reproduction. But what fuels this system, and what drives the evolution of such extreme apparent complexity? The principle underlying the answer to the first question was initially propounded by Ludwig Boltzmann, the nineteenth-century physicist and natural philosopher. Boltzmann had a tremendous admiration for Darwin and suggested, 'Available energy is the main object at stake in the struggle for existence and the evolution of the world'.

Thirty-six years later, Alfred Lotka (1922a, 1922b) interpreted Boltzmann's view to imply that available energy could be the central concept that unified physics and biology as a quantitative physical principle of evolution, stating, 'In accord with this observation is the principle that, in the struggle for existence, the advantage must go to those organisms whose energy-capturing devices are most efficient in

---

Box 1.1 Darwinism and Darwinian

Two of the most overused, and possibly abused, terms in the evolutionary lexicon. Although there exist notable exceptions (for example, Koonin, 2009; Koonin & Wolf, 2009), as with Lamarckism the precise meaning of the terminology can depend on the interpretation of an individual protagonist and may result in semantic quibbling. In this book the word is used in the strict contexts of variation and the process of natural selection as originally proposed by Darwin and Wallace. It is not, and cannot be, restricted to purely genetically driven evolution. Lamarckism is here defined as describing the inheritance of acquired characteristics, including somatic mutations. Because both Darwin and Lamarck predated the discovery of DNA, these definitions of Darwinism and Lamarckism here assume a more modern perspective.

---

directing available energy into channels favorable to the preservation of the species'. But, by themselves, such statements left unanswered the question of how different life forms could successfully reproduce themselves and also change. The question became, if available energy is the fuel, what is the directing force behind evolution? Although Boltzmann framed his comments in the context of Darwinian evolution (Box 1.1), his thoughts preceded the publication of Mendel's work on the genetics of peas. Consequently in the 1940s it fell to another physicist, Erwin Schrödinger, to combine Boltzmann's concept of available energy with that of a requirement for a heritable informational 'code-script' that specifies the form and function of all biological organisms. Life as we know it thus rests on the twin pillars of energy and information – consideration of both is essential for an adequate appreciation of the essence of life.

Today virtually all living organisms depend on deoxyribonucleic acid, DNA, as their primary source of genetic information – their 'code-script'. Some viruses utilise ribonucleic acid, RNA, a related nucleic acid, but these are very much the exception, and the amount of information encoded in RNA genomes, relative to that in DNA, is minute. So how did DNA achieve this dominant position? Why DNA – and not RNA? The current paradigm of information transfer in living organisms posits that the genetic information encoding proteins in

DNA is first copied into RNA, and the RNA copies are then translated into functional proteins by an elaborate molecular machinery (Box 1.2). But it was probably not always thus. In the initial stages of the evolution of life, it is believed that there was no DNA – it was an 'RNA world' – and possibly the appearance of RNA even preceded or was concomitant with the evolution of proteins themselves.

Both DNA and RNA are nucleic acids. They are both polymers consisting of a backbone of a long strand of alternating sugar and phosphate residues (Box 1.3).

An organic base is attached to each of the sugars. In RNA there are four principal types – adenine, cytosine, guanine and uracil. In DNA thymine replaces uracil. In any given chain of RNA or DNA the specific order of these bases in the polymer constitutes the genetic information. As famously pointed out by Watson and Crick, particular combinations of these bases have the ability to form complementary pairs with each other by forming hydrogen bonds. Adenine can pair with uracil or thymine, while guanine can pair with cytosine (Figure 3.1). This property allows two nucleic acid strands to base-pair with each other provided that the base sequence of one strand is complementary to that of its neighbour. It is the formation of this double helix that provides the fundamental basis for genetic inheritance. Today DNA exists almost exclusively in a double-stranded form. It is the 'double helix'. In contrast RNA molecules are predominantly single stranded, although this does not preclude the formation of double-stranded regions depending on base-pairing within a single strand. Chemically RNA and DNA differ principally in the nature of the sugar in backbone. The RNA sugar is ribose (hence RiboNucleic Acid), whilst that in DNA is deoxyribose (hence DeoxyriboNucleic Acid). This difference confers important different chemical and physical properties on the two nucleic acids, differ-ences that allow DNA to function more efficiently as a genetic information store.

By itself a description of the components of biological systems fails to capture the intrinsic nature of the process of life. The key to understanding life is that it is dynamic. It is not an ordered crystal structure but rather should be regarded as a system in constant flux where an elaborate organisation of multifarious chemical reactions is involved in maintaining individual organisms and enabling their

Box 1.2 Biological Information Transfer

The process of the transfer of genetic information in a biological system was initially formulated by Francis Crick in 1956 and subsequently published in 1958 (for an illuminating historical perspective, see Cobb, 2017): The Central Dogma. This states that once 'information' has passed into protein, it cannot get out again. In more detail, the transfer of information from nucleic acid to nucleic acid, or from nucleic acid to protein may be possible, but transfer from protein to protein, or from protein to nucleic acid is impossible. Information means here the precise determination of sequence, either of bases in the nucleic acid or of amino acid residues in the protein.

Although commonly interpreted as implying a unidirectional transfer of information from DNA to RNA to protein both Crick's text and the accompanying diagram in his notebook clearly envisaged, albeit cautiously, that in principle information could be transferred not only from DNA to RNA but from RNA to DNA.

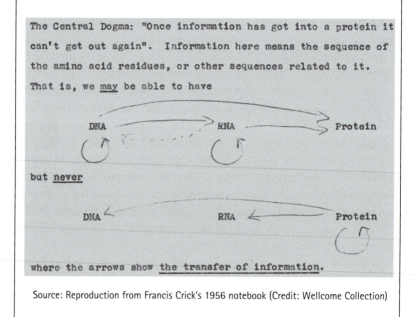

The Central Dogma: "Once information has got into a protein it can't get out again". Information here means the sequence of the amino acid residues, or other sequences related to it. That is, we may be able to have

DNA ⟶ RNA ⟶ Protein

but never

DNA RNA Protein

where the arrows show the transfer of information.

Source: Reproduction from Francis Crick's 1956 notebook (Credit: Wellcome Collection)

This caution was rewarded by the subsequent discovery that the RNA genome of certain viruses, the aptly named retroviruses, could be copied as DNA and inserted into nuclear DNA.

Box 1.3 DNA and RNA

The biological nucleic acids, DNA and RNA, are both polymers in which the monomeric units are nucleotides. Each nucleotide contains a 5-carbon sugar, usually ribose in RNA or deoxyribose in DNA, as well as a phosphate group and a heterocyclic nitrogenous base. The four bases most commonly found in RNA are adenine, cytosine, guanine and uracil, while in DNA thymine is found in place of uracil. Although the base-pairing rules – adenine pairs with uracil or thymine and guanine pairs with cytosine – are specific and conserved, modified variants of the bases often occur in both DNA and RNA. These include 5-methylcytosine, N6-methyl adenine, uracil, hydroxymethyluracil and glucosylated hydroxymethyluracil in double-stranded DNA as well as N6-methyladenine and other variants in transfer RNA.

---

replication for future generations. This concept of energy flux is again attributable to Lotka. Replication is dependent on the conservation of information within any biological system but, by the same logic, can allow small informational changes to determine changes in the characteristics of the chemical reactions and so direct the evolution of biological forms. In this context RNA initially had two essential attributes. Not only could it carry information in the form of a defined sequence of nucleotides, but it could also, by virtue of its chemical structure, catalyse a (rather restricted) set of chemical reactions. So by analogy to protein catalysts, aka enzymes, it could act itself as an enzyme, and indeed some RNA molecules still perform biologically crucial such functions in cells. Indeed the synthesis on the ribosome of the peptide bonds connecting individual amino acids in a protein chain is RNA-catalysed. That such an important feature of cellular information flow is still extant is one of the major indications that RNA catalysis is evolutionary ancient and likely developed in a world devoid of DNA.

In principle, given appropriate precursor chemicals, RNA can maintain itself by self-catalysed copying and processing. Such a process is likely inefficient. With the advent of the ability to synthesise protein molecules – even simple ones – the biological world would be transformed. Instead of the four basic structural units – the bases – in RNA, present-day proteins can contain up to at least

20 different fundamental units – as amino acids – in a polypeptide chain. Consequently not only is the repertoire of chemical reactions that can be catalysed by proteins very much greater than that of RNA molecules but also protein molecules either individually, or in collaboration with each other, can construct a scaffold that facilitates the close approaches of the chemical participants in a reaction. Or put another way, proteins can act to increase the local concentrations of reactants not only by possessing catalytic properties themselves – this is, after all, an essential attribute of enzymatic catalysis – but also by stabilising the structures of other catalytic macromolecules – be they RNA or protein – to effect the close spatial proximity of chemically reactive groups.

Ultimately all these chemical reactions that build an organism require energy (Lane, Allen, & Martin, 2010). But how can the process of life be reconciled with physical laws? At the heart of this question lies the apparent paradox first broached by the eminent physicist Erwin Schrödinger nearly 70 years ago. Like Boltzmann and Lotka before him, he addressed the fundamental issue of the nature of life itself as seen through the lens of a physicist. How is it that although Boltzmann's Second Law of Thermodynamics (Box 1.4) dictates that the universe ultimately approaches a state of maximum disorder, or entropy, that biological systems, as we observe them, appear to be, at the very least, minimising disorder or creating a system within themselves where order is actually increasing – or in Schrödinger's terminology – creating *negentropy* (see also Brillouin, 1953, 1956) The key to this apparent paradox is that biology operates in a thermodynamically 'open' system in which any local decrease or minimisation of entropy is more than compensated by an entropy increase elsewhere so that, on balance, overall entropy increases. In other words, because overall the system is heterogeneous, it is energetically possible for different parts of a connected system to gain or lose entropy, provided that there is no net loss of entropy. Indeed, because the energy required to reduce entropy locally is not utilised with 100% efficiency, there must overall be an increase in entropy. In this context what is important is Boltzmann's 'available energy' and not the overall energy content of a system. Energy is useless if it cannot be utilised. A tautology maybe but as Boltzmann put it in defining available energy:

Box 1.4 Second Law of Thermodynamics

The Second Law of Thermodynamics concerns processes that involve the transfer or conversion of heat energy. It posits that because such processes are not 100% efficient, some energy is 'wasted' and therefore a system progresses in the direction of increasing disorder, also known as **entropy**. This statement implies irreversibility and is the basis for the 'arrow of time'. There are many varied formulations of the Law. That of Planck states, 'Every process occurring in nature proceeds in the sense in which the sum of the entropies of all bodies taking part in the process is increased. In the limit, i.e., for [ideal] reversible processes, the sum of the entropies remains unchanged'. In the context of the Second Law, living organisms are never in states of thermodynamic equilibrium and must be considered as 'open' systems because they take in nutrients and give out waste products. A biological system is not reversible but operates within a non-equilibrium 'open' thermodynamic environment. Nevertheless, to satisfy the requirement of the Second Law that entropy increases as energy, there is, overall, a net increase in the system comprising an organism and the total environment in which it operates. A further crucial characteristic of the Second Law is that, as discussed in Chapter 2, it is statistical and thus probabilistic in nature. The Second Law has been, and to some extent still is, scientifically controversial, but as Arthur Eddington once said,

> The law that entropy always increases, holds, I think, the supreme position among the laws of Nature. If someone points out to you that your pet theory of the universe is in disagreement with Maxwell's equations – then so much the worse for Maxwell's equations. If it is found to be contradicted by observation – well, these experimentalists do bungle things sometimes. But if your theory is found to be against the Second Law of Thermodynamics I can give you no hope; there is nothing for it but to collapse in deepest humiliation.

(Although, for a dissenting view, see Hemmo and Shenker [2012].)

The general struggle for existence of animate beings is not a struggle for raw materials – these, for organisms, are air, water and soil, all abundantly available – nor for energy which exists in plenty in any body in the form of heat, but a struggle for [negative] entropy, which becomes available through the transition of energy from the hot sun to the cold earth.

Schrödinger (1944) points out very clearly that this statement is best appreciated in thermodynamic terms by invoking the technical definition of 'free energy' or, as he puts it, '[considering entropy alone' cannot account for [a biological system which feeds] on matter in the extremely well-ordered state of more or less complicated organic molecules]. But this balance differs between organisms. It is true for animals but much less so for plants and of course, both are part of the same complex system.

A fundamental question is then what is the nature of the molecular mechanisms that drive the creation of negentropy and the establishment of organisation? The creation of negentropy, otherwise a reduction in the intrinsic entropy of a system, implies increased order, a property of complex systems. But what do we understand by complexity? Complexity can be a slippery concept, distinct from, but related to, diversity but here is defined as a physical system containing a number of distinct components that interact directly or indirectly (Box 1.5). These components usually constitute a network. Diversity, a necessary condition for complexity, simply specifies the number of components without regard to their ability to interact. It is effectively a scalar property while by analogy complexity has some of the attributes of a vector. Although life is arguably one of the most extreme examples of a complex system, there are many other examples of such a phenomenon, for example, a galaxy cluster, an atom or even a city. In these cases available energy is used to create organisation in an open thermodynamic system. For a city the available energy would have been provided in early times by the agricultural harvesting of light energy and more latterly by fossil fuels.

The evolution of complexity in biological systems was also highlighted in Erwin Schrödinger's *What Is Life?* In his seminal lectures in 1943, he claimed that the laws of heredity required that the genetic material must contain a 'code-script' that determined 'the entire pattern

---

Box 1.5 Complexity and Diversity

A succinct and relevant definition of complexity and complex states has been provided by Chaisson (2015) – 'a state of intricacy, complication, variety or involvement among the networked, **interacting** parts of a system's structure and function; operationally, a measure of the **information** needed to describe a system, or of the rate of energy flowing through a system of given mass'. This definition emphasises not only that interactions between different parts of a system are a core component of complexity but also that the system can be described by information or by energy flux, as originally postulated by Lotka. For biological systems, at least, the intimate connection between information and complexity is discussed in Chapter 2. A related concept is that of 'complexification' (see Huxley in de Chardin, 1959), visualised as a process that is 'accompanied by an increase in energetic tension in the resultant corpuscular organisation, or individualised constructions of increased organisational complexity'. For further discussion of the concept, see Adami (2002; Adami, Ofria, & Collier, 2000).

Non-biological examples of complex systems share characteristics with biological systems. A simple example is an atom, composed of three different interacting sub-atomic particles – electrons, protons and neutrons – which like biological systems, form a dynamic structure. But this picture represents only one level of complexity. In the atomic nucleus both protons and neutrons are themselves made up of a mixture of three other sub-atomic particles – quarks. The precise flavour of the mixture determines whether the nucleon is a proton or a neutron. And again, the strong force between nucleons is mediated by apparently massless particles termed gluons. Like atoms, most complex systems can be considered to be a hierarchy of different levels of complexity.

Diversity, as the word implies, is a measure of difference and is shorthand for the number of distinguishable components in a system – for example, an ecosystem. Commonly used in the context of biodiversity as popularised by E. O. Wilson (1992), it is often broadly interpreted to include interactions between components and therefore complexity. Nevertheless, although it does not by itself imply interactions, diversity is the basis for complexity.

of the individual's future development and of its functioning in the mature state'. In other words, the individual's and by extension the biological system is specified by 'information'. At that time the nature of this code-script was obscure. Schrödinger appreciated that the encoding molecules must contain a non-repetitive molecular structure, but the experiments characterising the molecules that carried genetic information were only in their infancy – and in fact the first, from Oswald Avery and his collaborators implicating DNA as the responsible molecule, would not be published until the following year. Nevertheless, the concept that biological information is encoded chemically provided an essential link between the heritability and the thermodynamics of living biological systems. This mirrored Boltzmann's implied view that available energy could be the central concept that unified physics and biology as a quantitative physical principle of evolution.

This link is indeed central to the mechanism of the evolution of biological systems. In this context the overriding concept of natural selection, although usually discussed in relation to organisms, actually applies to the molecules that specify by their function the phenotype of an organism. At one level it is the thermodynamic characteristics of the molecules, particularly the macromolecules, that determine the properties of a biological system as a whole – however that system is defined. Most biological macromolecules, be they DNA, RNA or proteins, are long polymeric chains whose length is very much greater than their width. Superficially they may be compared to long lengths of flexible string such that the length of an individual molecule can be accommodated by many different pathways in three dimensions. It is this variety of pathways – more technically, configurations – that is one component of the intrinsic entropy – or disorder – of the molecule. Some of these long polymers, particularly RNA and proteins, perform structural roles. Others catalyse chemical reactions. Yet others combine these roles. But for those that catalyse chemical reactions, the close approach of the chemical groups required for catalysis within the molecule is essential. And because these groups have, in general, to be much closer to each other than the corresponding lengths along the backbone of the polymer only a few configurations out of a multitude will enable efficient catalysis. Selection for efficient reaction rate will thus inevitably result in a tightening of the frequency spectrum of the range of available configurations and so in a reduction of the intrinsic entropy.

On this view natural selection acts to confine macromolecules to particular preferred structures – it can act, for example, to reduce the wriggle rate of a moderately flexible polymeric string. In this context the formal description of such a string as a 'worm-like chain' is particularly apposite. This can be accomplished in several ways – by altering the sequence of the individual units in the polymer and so the immediate environment of the chemically reactive centre or, more frequently, stabilisation may be the result of two, or more, long polymers, binding to each other so reducing the wriggle rate of both. In this second situation, not only is the intrinsic entropy of both reduced but also the number of components involved is increased. Or, put another way, the *complexity* of the system is increased. Simple examples like this imply that natural selection can increase not only the apparent order in a biological system but also its complexity. To what extent this perspective can be generalised to encompass the huge diversity of the biological world that is apparent to us as an everyday phenomenon is a subject that will be explored later.

But for natural selection to act in this way, not only must a particular molecular variant be favoured for, say, an enhanced rate of a required reaction but also that variant must be replicated. It is here that the role of Schrödinger's 'code-script' is paramount. The 'code-script' defines and stores the information that specifies the sequence of macromolecules such as RNA and proteins. Initially in an RNA world RNA could perform both functions. It can act as an informational store and can replicate. But the information-carrying capacity of RNA, like that of the genetic code itself, has limitations. These limitations reflect the physical characteristics of the molecules involved. A simple analogy may be made with computing. In the early days of 'powerful' computers the information necessary to process experimental data – for example, generated by X-ray crystallography – was not stored first on a memory stick or even a hard disc – but instead on flimsy punched tapes. To transport the data from an X-ray run to calculations on the central computer on the Downing site in Cambridge required two people (Figure 1.1).

An improvement in technology led to the introduction of punched cards, but because in Cambridge the central university computer had reached the limits of its calculating capacity, the punched cards had to be stacked on shop trolleys and taken on the train to London to be

Figure 1.1 A low-tech method of information transfer. Punched computer tape – the output from an X-ray crystallographic study of the muscle protein, myoglobin – being transported outside The Arts School, Cambridge in 1958 by Bror Strandberg and Richard Dickerson.
Source: Courtesy of MRC-LMB archives and Peter Strandberg

processed on more powerful computers. These forms of information storage worked because there was nothing better, but in reality they were fragile, and the amount of information that could be stored on a particular structural unit – be it a card or a tape – was small. In the biological realm RNA is likewise more chemically fragile than DNA and, possibly because of their many different roles in information transfer, it is less feasible for RNA molecules to accommodate as much genetic information as the longer DNA molecules.

Another limitation of information storage is the ability to make optimal use of the potential informational capacity of a coding molecule. RNA molecules today contain predominantly four nucleotide bases, A, C, G and U. There are several scenarios for the evolution of such a heteropolymer, but in the simplest case it's possible to consider a primeval RNA molecule containing a heteropolymeric mix of all four bases. However, in such a molecule the ability to read a sequence to form a protein then depends on the stability of the interactions of the decoding reading heads with the base sequence itself. Initially only the

most stable interactions would likely be accessible, leaving other information potential in the RNA molecule essentially unreadable, or effectively nonsense. But as more and more stable reading head interactions
became accessible, so the amount of nonsense would decrease. Put
another way, the information storage in an RNA molecule would
become more efficient, and the predictability of the reading process
would increase. This is directly related to a quantity termed Shannon
entropy, a central tenet of information theory put forward by Claude
Shannon in 1948. In this theory an increase in predictability decreases
Shannon entropy. As foreseen and paraphrased by Gilbert Lewis, an
eminent chemist and discoverer of the covalent bond, discussing chemical entropy: Gain in entropy always means loss of information, and
nothing more.

There are further parallels with the nature of information encoded
in DNA with that encoded by computers. As pointed out by John von
Neumann (1958), in computing there are two modes of specifying
information – analogue and digital. In the analogue mode the information is expressed in principle as a continuous variable – the classic
clock face is an everyday example, while Charles Babbage's Analytical
Machine, one of the first real computers, used an arguably analogue
mode. In contrast digital information is stored in an essentially dis-
continuous fashion. Again digital clocks are an excellent everyday
example, and indeed most modern computers store information in a
digital mode. DNA, however, combines analogue and digital modes
(Marr, 2008) and encodes different types of information in each mode.
For example, the encoding of codons specifying protein and RNA
molecules is purely digital, while the specification of the physical
properties of DNA – how easily it bends or the strands separate – is
the ensemble of the physicochemical properties of a more extended
DNA sequence. Simplistically the digital information determines the
nature of the macromolecules in a cell, while the analogue information
determines the regulation of the production of the macromolecules.
However because both the analogue and digital information are
encoded by the sequence of four bases in DNA, they are not independent, especially, for example, where the specifying sequences overlap.
Arguably it is the ability of DNA to combine these two types of
information that enables it to act as such an efficient genetic repository. There is another biological information processing system that

combines analogue and digital characteristics. The mathematician John von Neumann pointed out that within a single neuron a nerve impulse is essentially digital in character. A neuron either fires or it doesn't. However the transmission of the impulse signal to a neighbouring nerve cell via a synapse depends on the concentrations of certain small molecules – neurotransmitters – in the small gap between the neurons. In this case the transmission has an analogue character because concentration is a continuous variable.

Although combining analogue and digital information in the same molecule may place restrictions on the usage of both components, the use of both modes in principle increases the information carrying capacity of the molecule. If you like, the DNA cable has two channels rather than one, and so again, in a biological system, the utilisable information is increased. In his original proposal Schrödinger framed his argument in terms of classical thermodynamics – that is, the entropy in his definition referred to the classical Boltzmann entropy. However, importantly, the information content of the code-script is a direct reflection of the complexity of the operating system. The more information that is encoded, the better the system is defined. Or put another way, biological mechanisms are selected so that, as far as possible, nothing is left to chance within the physical constraints acting on the system. An accumulation of information in this way thus implies a lower Boltzmann entropy or 'negentropy', and the organisation of information within the 'code-script' itself contributes to negentropy. The nature of the information in the 'code-script' – be it RNA or DNA – is a primary vindication of Schrödinger's hypothesis.

The Transition from an RNA World to a DNA World: A Major Increase in Complexity and Information

The supplanting of RNA by DNA as the major information store in biological systems is a prime example of an increase in complexity. However in present-day biological systems, RNA still acts as a messenger molecule (mRNA) for directing the synthesis of proteins, still in the form of tRNA acts as an adaptor for decoding the nucleotide sequence in mRNA into a specific amino acid sequence in proteins, and still in the ribosome acts as a structural and catalytic component of information flow. RNA molecules, in different forms, have thus retained, not necessarily in their original guise, many of the essential features of information flow that would have been essential in a biological world lacking DNA. Conversely, because DNA contains deoxyribose in place of ribose in its sugar-phosphate backbone, the potential of a DNA molecule for catalysis relative to RNA is much diminished. What cellular RNA molecules – strictly RNA molecules encoded by DNA – have lost is the ability to act as a template directly for their own replication. As the molecule conferring heritable properties DNA, it is the DNA sequences that are replicated so that any heritable changes – mutations – in the sequence can be passed on to future generations.

The consequences of the increase in nucleic acid complexity from RNA only to RNA plus DNA would have been a game-changer. In essence the existence of two fundamental types of nucleic acids with different functions would require, as an absolute necessity, some form of communication – interaction – between them. Not only would the sequences of the different RNA molecules have to be encoded in the DNA but also because within a cell the abundance of specific RNA molecules can differ by up to several thousand-fold the DNA must contain information specifying the number of RNA molecules to be synthesised. For example in the simple bacterium *Escherichia coli*, there may be of the order of 20,000 of each of the three RNA components of the ribosome, while some mRNA molecules may be present in less than 10 molecules per bacterium. Indeed, some mRNA molecules may only be synthesised when the organism is exposed to very different environmental circumstances. So not only must the DNA contain the information specifying the quantity of each RNA molecule required by the cell, but it could also contain information for varying – regulating – the quantities required depending on the prevailing internal and external environmental conditions.

In this DNA world DNA and RNA molecules are not stand-alone entities. A network of interactions is essential for the efficient operation of information flow. This is not simply an increase in the diversity of the different nucleic acid molecules. It is an increase in the complexity of the biological system. One consequence of an increase in complexity is the possibility of the development of new functional modes that were precluded in a simpler system. Such novel 'emergent properties' reflect an increasing self-organisation. In a transition from

RNA alone to RNA plus DNA any constraints imposed by the limitations of an RNA genome might by relieved by a separation of function. One such possible example of an emergent function is the ability of DNA molecules to adopt a higher energy state in which the long DNA chain is coiled – known as a supercoil because the DNA double helix is itself a coil. In this form the DNA stores energy, is more easily packaged into a small volume and creates a situation in which the activity of one gene on a DNA molecule can influence the activity of its neighbour. Supercoiling requires that the rotation of a DNA molecule about its length be restricted – most easily accomplished by having a circular DNA molecule or attaching the two ends of a linear DNA molecule to a fixed scaffold. The dual roles of DNA as an information repository and an energy store uniquely represent the twin underpinnings of a low-entropy complex system – information coupled with efficient energy utilisation.

More importantly, not only does DNA store information but that information is mutable, enabling modification of the phenotype and selection for more adapted characteristics of an organism in a changed environment. Critically the 'code-script' is also highly dynamic. The ability of complementary DNA strands to pair with each other over limited regions not only facilitates the shuffling of genes during the formation of germ cells but also enables short DNA segments – so-called jumping genes or mobile DNA elements – to move from place to place in the DNA genome, essentially acting as an intrinsic mutagen. Indeed, the ubiquity of jumping DNA implies that it is maintained by strong selective forces. These could reflect simply the presumed 'selfish' character of such DNA elements or, alternatively, the ability of such elements to create mutations in the genome that could act as an evolutionary driver.

The acquisition of emergent properties by an increase in complexity from an RNA-only world does not necessarily imply that a double-stranded RNA genome able, for example, to support supercoiling could not exist and that a different means of regulating information flow with RNA alone is not theoretically possible. As with the evolution of organisms, the evolution of biological molecular systems depends on previous history as well as circumstance. In this sense it is said to be Bayesian (Box 1.6) in character. Perhaps the selective advantages conferred by DNA would, if reproduced in RNA, be similarly effective.

Box 1.6 Bayesian Logic

Thomas Bayes (c. 1701–1761) was an English clergyman who took a deep interest in probability theory. His principal accomplishment – published posthumously – is known as Bayes' theorem and describes the probability of an event based on prior knowledge of conditions that might be related to the event. The theorem was subsequently refined and popularised by the French mathematician Pierre-Simon Laplace (1749–1827). Biological evolution perfectly exemplifies Bayesian logic because by changing – usually substantially increasing – the probabilities of particular interactions at the expense of others so that the probable pathways for future evolution are severely restricted. In this context Dawkins' illustrative example of the implausibility of the appearance of 'crocoducks' is well worth reading (Dawkins, 2009). The common ancestor of crocodiles and ducks substantially predates both, precluding the evolution of the one into the other.

That is, without the appearance of DNA, the functionality of double-stranded RNA might have evolved to adopt roles analogous to those inherent in DNA. Whatever the earliest steps in evolution, once DNA became established as a separate functioning entity, its apparent intrinsic advantages as a genomic information repository would lead to its present dominating position.

Information and Complexity

By recognising the informational role of the 'code-script', Schrödinger coupled the concepts of the information contained within a DNA sequence and the complexity of living systems. But, as an informational carrier, DNA is restricted by genetic transmission and arguably is not the sole source of information that currently drives the evolution of some complex biological systems. Complex mammalian societies – not just human ones – can, and do, transmit information utilising 'cultural' modes that do not directly require DNA transmission (Chaisson, 2013). Indeed, it can be argued that the complexity of a society or polity is dependent on the amount of cultural information available (Tainter, 1988). Are the physical consequences of this form of information transmission related to the 'negentropy' postulated by Schrödinger? If so, the implication is that the development of complexity in biological systems is directed primarily by information input and is itself adapting to inputs from different sources. Put another way, the physical phenomenon of increasing complexity in biology represents the evolution of a seamless transition from a purely DNA-based information system to one utilising additional sources of information. Such evolution is naturally conditional on an adequate supply of energy.

Throughout the passage of time biological diversity and complexity have increased with only transient interruptions occasioned by geological or astronomical interventions (Bonner, 1988; Wilson, 1992). This observation raises two related issues. Not only does the increasing immensity of the complexity that we now perceive appear inherently improbable (or even to some implausible), but also why is the process apparently directional? An illuminating perspective on the aspect of probability was provided by Gilbert Lewis (1926) again:

> Borel makes the amusing supposition of a million monkeys allowed to play upon the keys of a million typewriters. What is the chance that this wanton activity should reproduce exactly all of the volumes which are contained in the library of the British Museum? It certainly is not a large chance, but it may be roughly calculated, and proves in fact to be considerably larger than the chance that a mixture of oxygen and nitrogen will separate into the two pure constituents. After we have learned to estimate such minute chances, and after we have overcome our fear of numbers which are very much larger or very much smaller than those ordinarily employed, we might proceed to calculate the chance of still more extraordinary occurrences, and even have the boldness to regard the living cell as a result of random arrangement and rearrangement of its atoms. However, we cannot but feel that this would be carrying extrapolation too far. This feeling is due not merely to a recognition of the enormous complexity of living tissue but to the conviction that the whole trend of life, the whole process of building up more and more diverse and complex structures, which we call evolution, is the very opposite of that which we might expect from the laws of chance.

But if an evolving complex system is improbable, is an even more complex system even more improbable? The answer to this question is likely no, a conclusion rooted in fundamental thermodynamic properties of complex systems. The biological world we know is a superb example of a highly efficient mechanism for utilising its principal energy source – sunlight. In this context it is representative of an organised low entropy system associated with high exploitability and control of energy. Conversely a high entropy system is associated with low exploitability. Therefore natural Darwinian selection will inevitably select for more complex and information-rich systems with more regulation provided that there is sufficient energy available. Of course if energy is limiting, so also is the ultimate complexity attained. But within these limits the evolution of biological complexity becomes a positive feedback process and is consequently directional.

2

The Nature of Biological Information

Information, Complexity and Entropy

... the true logic for this world is the calculus of probabilities ...
—James Clerk Maxwell, 1850

Le hasard pur, absolument libre et aveugle, est la véritable racine du
merveilleux édifice de l'évolution.
—Jacques Monod, 1970

We have examined the 'biological phenomena' of a nonliving device
and have seen that it generates exactly that quantity of entropy which
is required by thermodynamics.
—Leo Szilard, 1929

Information is physical.
—Rolf Landauer, 1991

Schrödinger's solution to the question, 'What is life?' was to postulate that 'information' in the genetic material, characterised as the 'codescript', was essential for the thermodynamic minimisation of entropy (Box 1.3) accompanying the increasing organisation of biological systems. This critical insight paved the path for understanding the mechanisms of biological evolution. But how would a 'codescript' potentially lead to an increase in organisation? A simple, although not exact, analogy would be a small person given a pile of (appropriate) Lego bricks and asked to build a model of a space rocket. With no knowledge of how the rocket is put together, this would be an essentially impossible, or at the very least highly improbable, task, akin to the thousand monkey scenario (Chapter 1). However, given a book of instructions, such small people would – and, by observation, do – make short shrift of the problem. These instructions thus facilitate the construction of a specific complex structure and in so doing dramatically increase the probability of success. Similarly, without appropriate instructions 'all the king's horses and all the king's men' failed to put the smashed Humpty Dumpty together again. Again they lacked the information to transform an essentially irreversible event – with its statistically highly improbable reversal – into an organised structure (for an excellent discussion on this point see Penrose [2010]). To take a biological example, what is the probability of a living bacterial *Escherichia coli* cell spontaneously arising due to a thermal fluctuation in an equilibrium system? Using simple assumptions, a calculated value is (Morowitz, 1978):

$$10^{-10^{11}}.$$

A very small number indeed, corresponding to an entropy difference of about 3.2×10^{-12} JK^{-1} between an *E. coli* cell and its constituent matter in the equilibrium state (Box 2.1). However, using the instructions in the DNA codescript, it may take only 20' for an *E. coli* cell to reproduce itself.

Box 2.1 Scientific Units Relating to Entropy

Entropy: entropy is given the symbol S, and standard entropy, measured at 298 K and a pressure of 1 bar (approximately 1 atmosphere), is given the symbol $S°$. The units of entropy are JK^{-1} or joules/absolute temperature (Kelvin). A joule is a unit of energy and is equal to 10^7 ergs (used in Figure 8.1). ΔS represents the change in entropy for a given physical or chemical transformation.

The Boltzmann constant, given by the symbol k or k_B, is the proportionality factor that relates the average relative kinetic energy of particles in a gas with the thermodynamic temperature of the gas. It has the same dimensions as entropy and relates the statistical definition of entropy to the classical definition and to Shannon entropy.

Such examples imply that information is an intimate adjunct to what is termed biological complexity. But the simple analogy of DNA or a genome to an instruction book by itself lacks a crucial property of the DNA codescript. DNA information can be changed – it is mutable. It is this property that enables the operation of Darwinian natural selection and the consequent evolution of complexity.

By its very nature, complexity requires diversity – heterogeneity – so any mechanism that always produced perfect carbon copies would not in itself generate complexity. In simple terms complexity – be it, for example, a structure of many interconnecting parts or a communication system between different entities – represents one or possibly only a very few – of the many possible combinations of interconnections between the component parts. Put another way, information, by increasing the probability of particular highly unlikely events, can result in the emergence of complex system with a high degree of organisation. In thermodynamic terms this implies a decrease in entropy. However, the second law of thermodynamics posits that when work is done in any isolated system, there is a natural tendency for such a system as a whole to degenerate into a more disordered state – or in other words for entropy to increase (Box 1.3).

The nature of information itself is the key to this apparent contradiction between the second law and the evolution of complex systems. The importance of information in relation to thermodynamics was first realised by James Clark Maxwell in the nineteenth century. Maxwell argued that if an intelligent being – later conferred with a demoniacal, although not a malicious, persona by William Thomson (Lord Kelvin) (1879) – when presented with a contained mixture of fast hot gas particles and slow cold ones and possessing information about the velocities and positions of the particles in a gas, could transfer the fast, hot particles from a cold reservoir to a hot one and vice versa, in apparent violation of the second law (Figure 2.1a). Maxwell's demon revealed the relationship between entropy and information for the first time – demonstrating that, by using information, one can apparently relax the restrictions imposed by the second law on the energy exchanged between a system and its surroundings.

In this context the small boy with his instructions building a model of a space rocket is only an example – albeit a rather more sophisticated one – of Maxwell's original demon (Figure 2.1b).

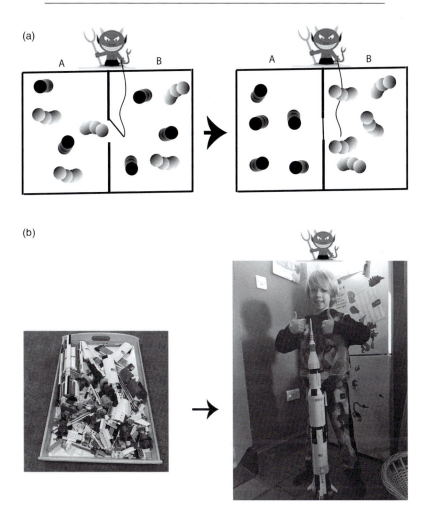

The Force (Maxwellian) Awakens

Figure 2.1 (a) Maxwell's thought experiment. The demon 'knows' which gas molecules are hot (grey) or cold (black) and by allowing only cold molecules to move to the left through the door and hot molecules to move to the right effects a separation into two homogeneous compartments, A and B. Source: Figure redrawn from Wikipedia.org and reproduced under the terms of the Creative Commons Attribution-Share Alike 3.0 Unported licence. (b) The importance of instructions. A pile of Lego bricks transformed into a model of a rocket by a small boy (the author's grandson). Maxwell's demon approves. *A black-and-white version of this figure will appear in some formats. For the colour version, refer to the plate section.* Source: Photos by H. Travers and J. Burt

Furthermore Lego instructions have one other characteristic in common with biological systems – they are modular (Box 2.2). The translation of a messenger RNA to protein by ribosomes and tRNA fall into both categories.

Conceptually information is often considered to be essentially an abstract phenomenon. But such a perception disguises its real nature. As Landauer once pithily put it, 'Information is physical' (Landauer, 1991), and because it is physical, work is required for its creation. In this context Schrödinger's formulation was prescient. The genetic 'codescript' is both a code, i.e., it imparts information, and also a script – it is written, in this case as a sequence of bases joined together by a chemical backbone. There are, of course, many other examples of information repositories – inscriptions on a page or stone (writing and printing), electronic memories and, again in the biological sphere, the pattern of neural interconnections that constitute memory. All of these require energy input for their creation, and all, at least potentially, can promote an increase in complexity.

The demon's apparent contradiction between the second law and information is consequently an illusion – and it was never Maxwell's intention to vitiate the second law (Box 2.3). By its very nature the second law is probabilistic, and ultimately information, which can alter the probability of the outcome of a process, is a key consideration in its operation. The contradiction depends on the term 'isolated system'. As devised Maxwell's demon operates on an isolated system but in doing so interacts with it, although it is itself not part of it yet imparts information to it. Similarly the biological world is not, by itself, an isolated system. It is an a very small component of a much larger universal energetic system on which it feeds to acquire the energy necessary to maintain and evolve its characteristic information-intensive complexity. The universal system is bound by the second law – entropy increases to a maximum – but that does not preclude the formation of isolated pockets of regions of entropy minimisation (or 'negentropy' in Brillouin's [1953] and Schrödinger's terminology [1944]). The biological world operates in such an 'open' (non-equilibrium) environment. That is, the work required for the maintenance of biological complexity results in an increase in entropy, which is manifested in the universal system but not in the biological system as narrowly defined. Such a system is also directional. Being far from equilibrium, such systems can

---

Box 2.2 Modularity and Its Role in Evolution

Modularity is a core attribute of complex biological systems and describes a mode of organisation in which specific functions are performed by discrete (although linked) dedicated structural units, each of which comprises a module. The importance of modularity as a vehicle for optimising energetic efficiency is strikingly illustrated in the parable of the two watchmakers Hora and Tempus (a creation of the economist Herbert Simon [1962], later popularised in this context by Smith & Morowitz [2016]). Both made very fine desirable watches, but while Hora prospered Tempus fell into penury. The important difference between them was that Hora manufactured his watches on a modular principle, whereas Tempus only built a single watch at a time and had to maintain its structure while he did so. The result was that when, for whatever reason, manufacturing was interrupted, afterwards Hora could continue construction by assembling pre-existing modules, whereas Tempus had to start again from scratch. Hora's energy efficiency (number of completed watches / time invested) was consequently much greater than that of Tempus. The corollary is that Hora's success depended on constancy of design. Exactly like a Lego model the modules were assembled in a definite order, a requirement that meant the intermodular interactions were rigorously specified. Similar principles apply to biological evolution. A primitive chemical trait can likely become irreversibly embedded as an enduring characteristic. One such example is the modular organisation of many eukaryotic genes. Another is the Na+/K+ external/internal ratio current in all forms of life (Chapter 7). To change form an initial ionic equivalence to the current concentration imbalance would likely necessitate a ratio change of 2–3 orders of magnitude. Given the modular nature of cells and pre-cells, it seems implausible that such a change could be driven simply by natural selection. More likely biological evolution has a Horaesque quality dominated by Bayesian logic. Such a property imparts directionality – although it does not imply imposed direction. It can also lead to evolutionary cul-de-sacs, where the co-evolution of distinct organisms generates complete symbiotic dependence – for example, figs and fig-wasps.

Box 2.3 Entropy and Information

Entropy is a fundamental concept of energy and, as frequently under-stood, is a measure of the disorder in an isolated system. Any chemical reactions that sustain such an isolated system, such as a human, are inefficient (<100% utilisation of available energy), and the energy not used contributes to disorder. The consequence is that entropy increases with time, aka time's arrow. The complexity of the biological world we observe every day is maybe counter-intuitive to this concept. Yet con-sider the example of the Lego model. The number of arrangements of the pile of Lego bricks, or even the number of possible constructions that could be built from them, is very large, yet there is in principle only one perfect solution for building a model of the rocket. The number of disordered arrangements represents a high-entropy state, but the model of the rocket is an ordered low-entropy state. It is the information in the instruction book that drives the local reduction of entropy. And so it is with DNA.

---

accommodate apparently improbable storage of energy in high-energy electronic states.

A crucial attribute of 'information' is the requirement for a decoding receiver, allowing the discrimination of a usable signal from untrans-latable noise. Consider a crowded room full of many conversations. As their number increases, the 'noise' level also goes up, and the ability to filter a particular conversation from the many in the room goes down. As a result, the certainty – or probability – of 'information' transfer in a particular conversation is concomitantly decreased. In the 1940s this phenomenon was addressed mathematically in a more general form by Claude Shannon studying the efficiency of information communication (Shannon, 1948). He introduced the concept of what is now known as Shannon entropy (a term controversially suggested by the computer scientist John von Neumann), which he demonstrated to be equivalent to a shortage of information content in the message. In other words, the larger the Shannon entropy, the greater the uncertainty of information transfer to the receiver. Mathematically the formulation of Shannon entropy is very similar to that of thermodynamic entropy (Box 2.4), raising the question as to whether these two quantities are connected. Such a connection, of course, would be directly relevant to biological

Box 2.4 Shannon and Boltzmann Entropies

Information in the codescript can be described by **Shannon entropy** (H)

$$H = -\sum_i p_i \log{_b} p_i.$$

Entropy of the biological 'system' (however defined) is described by **Boltzmann entropy** (S) (in a homogeneous system) or more generally by **Gibbs** entropy

$$S = -k_B \sum_i p_i \ln p_i,$$

where k_B is the Boltzmann constant. These equations are similar in form, and both express uncertainty, either in encoded information (Shannon) or physically (Boltzmann).

However, caution is in order – Shannon and Gibbs–Boltzmann entropies are not strictly equivalent (Nigatu, Henkel, Sobetzki, & Muskhelishvili, 2016). In typical pithy style, Jaynes (2003) commented, 'We must warn at the outset that the major occupational disease of the field is a persistent failure to distinguish between informational entropy [Shannon entropy, H function] and experimental entropy [Gibbs–Boltzmann entropy] of thermodynamics. They should never have been called by same name: the experimental entropy makes no reference to any probability distribution, and the informational entropy makes no reference to thermodynamics. Many textbooks and research papers are flawed fatally by the author's failure to distinguish between these entirely different things, and in consequence proving nonsense theorems. But, [to confuse matters further,] in the case where the problem happens to be one of thermodynamics there is a relation between them.'

To take a simple example. The sequence, or signal, in a DNA molecule can be analysed by Shannon's rules in terms of available combinatorial possibilities. These depend on the base content and in a random sequence are maximal (unless there is significant sequence bias) at 50% A/T. Accordingly, Shannon entropy is also maximal at this point (Schneider, 1997; Nigatu et al., 2016) because the number of sequence possibilities is greatest, and thus the probability (or certainty) of the occurrence of a particular sequence is a minimum. By contrast, the physical properties of DNA, which can be expressed in thermodynamic terms and are sequence dependent, exhibit a different dependence on base composition (see Chapter 3).

systems in which information is directly encoded by the 'codescript'. In simple physical systems, Shannon entropy is often loosely taken to be equivalent to thermodynamic entropy, and indeed, even in certain non-equilibrium situations, it can be shown that Shannon entropy, by determining the thermodynamic energetics, has a clear physical meaning. In biology the existence of multiple pathways of information flow from DNA to the observed outcome currently precludes such a conclusion (but see next).

Shannon and Communication

At the time he wrote his classic paper, Claude Shannon was working as an electrical engineer employed by the Bell Telephone Laboratories, and his work focused on the practical issue of how best to encode the information a sender wishes to transmit. That is, his analysis was concerned with the nature of the signal and the ability of, for example, a cable to transmit 'information' rather than the outcome consequent upon decoding (Figure 2.2). In telecommunications the decoder is designed to be compatible with the signal generator or encoder, but, as especially also in biology, decoders can be selective, being receptive to only a relevant fraction of the transmitted signal (think of a radio tuning into different radio frequencies). For this reason, it is important to distinguish between the statistical properties of the signal – loosely described by informational, or Shannon, entropy – and the physical outcome of the interaction of the signal with the decoder, which can be described by thermodynamic parameters. Of course the signal itself can also be described in physical parameters – for example, as a component of the electromagnetic spectrum – but its ultimate physical effect within a complex system depends on decoding.

The importance of a decoder, or interpreter, is very relevant to biology. Without a decoder a sequence of base-pairs coding for a protein

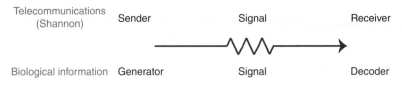

Figure 2.2 Pathway of information communication.

has no digital meaning. It is only, for example, when such a nucleic acid sequence is translated into a protein sequence by the ribosome acting in concert with tRNAs does that particular 'information' it encodes enter the physical world. Without an appropriate context, any meaning is unintelligible – although computer programmes can do a good job of decoding albeit with crucial added contextual information. The importance of context is underlined by the natural, albeit small, variation of the genetic code. It is often stated that the genetic code is universal. It is not, even though there is a dominant version. Some triplet codons encode different amino acids in different decoding systems. These systems can be in organisms or organelles, such as mitochondria. One such codon is UGA, which in different contexts can encode protein chain termination with no decoding tRNA (a STOP codon) or one of tryptophan, cysteine, glycine or selenocysteine. For example, consider the coding sequence:

> . . .GGU GCA AUU UGA CCG AAA ACG UAA CTA. . .

This RNA sequence translates to a protein sequence as:

> . . .gly ala ileu STOP . . . .   in plants and animals
> . . .gly ala ileu trp lys thr STOP . . . . in yeast mitochondria
> . . .gly ala ileu cys lys thr STOP . . .   in *Condylostoma magnum*
> (a ciliate)
>
> . . .gly ala ileu gly lys thr STOP . . . . in 'microbial dark matter'
> OR . . .gly ala ileu sel lys thr STOP. . . . in many organisms but with
> additional 2° structure
> information for selenocysteine
> (sel) in the mRNA

In this example, decoding requires the formation of a chemical complex mediated by base-pairing between the mRNA base sequence and a tRNA. In the case of a STOP codon, a protein release factor senses a lack of tRNA and binds to the ribosome. For both, decoding requires the chemical interaction of two molecular entities. The mRNA–tRNA interaction requires that **both** partners contain the appropriate complementary base sequences. Is the tRNA equivalent to a 'knowing' Maxwell's demon, or does the mRNA–tRNA recognition complex constitute an energetic equivalent of 'information'?

This example of variation among decoders implies that the expression of the signal into a physical outcome is not necessarily intrinsic to the signal, in this case the mRNA sequence, but requires inter-action with the decoder. In other words, as importantly pointed out by Szilard, 'information' is associated with a binary process. This property is intrinsic to its nature (Muskhelishvili, 2015). In biological systems, decoders are often highly selective in the same way that a radio can be tuned to receive only, usually, a narrow bandwidth of frequencies (a channel). The ubiquitous hawkweed and dandelion flowers provide another example of such selection. Humans and bees perceive these flowers in different ways. In this case consider the 'signal' as the spectrum of reflected photons, which extends from the usual visual range into the ultraviolet (UV). However, because the human decoder is insensitive to UV photons – it is selective – humans only perceive a hawkweed or dandelion as a yellow blob. In contrast, bees have sensors for UV light and can discriminate between the outer petals and the centre, which is the principal source of the reflected UV light (Figure 2.3). To a bee this is extremely useful because the centre is the source of pollen and nectar. Interestingly, the ability of flowers to generate 'blue haloes' of reflected blue-UV light depends on a microscopic pattern of ridges on the surface of a petal, a structure that likely evolved independently several times

Visible   UV

Figure 2.3 A hawkweed flower visualised with light of different wavelength, as perceived by a human (left) and a bee (right). *A black-and-white version of this figure will appear in some formats. For the colour version, refer to the plate section.* Source: Reproduced from Wikipedia.org under the terms of the Creative Commons Attribution-Share Alike 3.0 Unported licence

(Moyroud et al., 2017). Bees and humans thus perceive distinguishable, albeit overlapping, signal channels.

An analogous principle of selective decoding also operates in language (Barbieri, 2016). In English the word 'ape' could conjure up a vision of a King Kong-like monkey. In contrast, the same word in Italian describes a small buzzing insect – a bee. However, in French there is no such word. To a Frenchman 'ape' is meaningless – it is not contained in the French decoder. Taken to its extreme, this implies that some signals cannot be decoded because no current decoder exists. This is true of Linear A, an ancient Minoan language, but not of its successor, the Mycenean Linear B, which was only successfully deciphered in the twentieth century. Another example is the acquisition of information from a book by a blind person. Here it depends on the nature of the signal. The printed word, in contrast to a Braille-encoded word, conveys no information to the blind recipient because it cannot be received.

Landauer's Principle: Information Is Physical

The concept that information acquisition, or measurement, has a thermodynamic cost was first advanced in 1929 by Leo Szilard, who, commenting on Maxwell's demon, observed that the demon's 'measurement procedure is fundamentally associated with a definite entropy production'. He further identified the 'fundamental amount' of entropy associated with 'information' to be $kT\ln 2$ (where k is Boltzmann's constant and T the absolute temperature). This concept received experimental confirmation in 2014 (Koski, Maisi, Pekola, & Averin, 2014). In 1961 Rolf Landauer extended this idea to consider a simple ON/OFF gate in a computer. In such an ideal system, switching from the ON state to the OFF state discards the contained information in an irreversible manner (Figure 2.4). Because the ON state contains a single information bit (Box 2.5), this process releases entropy with the minimum value of $kT\ln 2$. Landauer argued that in a computer the OFF state is ill-defined, and therefore to reset to ON state, additional information from an outside source – for example from a computer program – is required. This resetting restores a single informational bit to the gate but requires work. In this cycle process energy is both released and added – the basis for Landauer's aphorism that 'information is physical'.

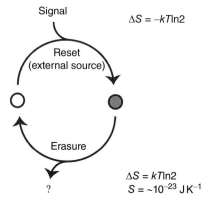

Resetting one bit

Figure 2.4 Landauer's switch.

Box 2.5 Bits and Bytes

A bit (binary digit) is the fundamental unit of information storage and is denoted by just a 0 or 1 in binary code. A byte is 8 bits, while a megabyte is 1,024 or 2^{10} bytes. The term 'bit' is derived from the notion of information as associated with a binary decision process – e.g., ON or OFF (Szilard, 1929). A binary code would thus have the form . . .OOIOIOOIII. . ., etc. The ln2 component of the minimum entropy associated with a bit, $k$ln2, is a reflection of the binary process. However, because the binary process involves a binary interaction, $k$ln2 is a measure of the thermodynamic entropy associated with an 'informative interaction' (see following).

Landauer's theory suggested a powerful reason why Maxwell's demon could not violate the second law. The demon would need to erase ('forget') the information it used to select the molecules after each operation, and this would release heat and increase entropy, more than counterbalancing the entropy lost by the demon. Put another way, erasure carries a thermodynamic penalty. Maxwell's proposal was simply a thought experiment, but recently experimental proof of Landauer's principle that information can be converted to energy has been demonstrated by elegant physical techniques (Toyabe, Sagawa,

Ueda, Muneyuki, & Sano, 2010; Bérut et al., 2012; Parrondo, Horowitz, & Sagawa, 2015). Convincingly, the minimum observed energy produced by erasing one bit closely approached Landauer's prediction.

Landauer considered a very simple system – basically a logical (electronic) gate that could exist in two states and could cycle between these states. The states, by themselves, were not obviously involved in any external interactions, yet, importantly, required resetting by a signal from a computer. But is this model applicable to biological information?

A possible analogy would be the reception of a signal by a biological sensor. The signal could, for example, be a small molecule that would bind to the sensor forming a binary complex (Figure 2.5). This could be considered as an 'informative interaction'. However, unlike computer switches, molecular complexes of this type have only a finite lifetime and after a while dissociate, freeing the sensor for a subsequent interaction and resulting in an increase in thermodynamic entropy. Formally this signal/sensor biological cycle is analogous to Landauer's ON/OFF switch with dissociation corresponding to erasure and association to resetting from an external computer.

Is such a model also applicable to DNA? The actual sequence of bases in a nucleic acid – whether RNA or DNA – is just that – a sequence.

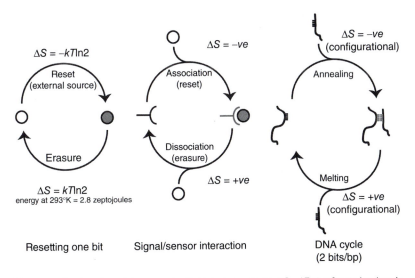

Figure 2.5 Landauer's model compared with biological systems. $\Delta S = kT\ln2$ refers to Landauer's model (left). The values for the other processes depicted are uncertain.

Without a receiver – a nucleic acid strand with a complementary sequence or maybe a protein – it is equivalent in informational terms to a string of Linear A characters being presented to a naïve reader. By itself the sequence is defined, and therefore each base is equivalent to 1 bit of information. But, in the context of biological information, the fundamental functional digital informational unit in biology is likely the base-pair. Any nucleic acid sequences with the potential for base-pairing also have the property of cycling between two distinct states. Annealing generates a base-paired double helix, while melting once again separates the two strands (Figure 2.5). Can cyclic annealing and melting be compared to Landauer's ON/OFF gate cycle? In Landauer's cycle, entropy is released when information is obliterated and reduced on resetting. To that extent this process is consistent with Schrödinger's proposition that information input can be correlated with 'negentropy'. But is this also the case for a single-strand/double-strand nucleic acid cycle? Here the base-pairing of a sequence of bases (as in Figure 2.5) can bring together, for example, two single strands. The formation of this complex generates a single molecular entity with different physical properties from its components. The complex can occupy a smaller molecular volume, but also, because it is larger, its kinetic properties will be different. Individually the two component strands will be very flexible and so have a lot of wriggle room (or configurational space) because the individual chains have the ability to bend in lots of different directions. However when a short region between the two is base-paired, the wriggle room is constrained – there is now only one molecular entity, and in the base-paired region it is stiffer. In thermodynamic terms base-pairing of the two molecules reduces the configurational entropy, while melting acts in the opposite direction. In other words, the information in a base-pair decreases the intrinsic entropy of the system – exactly in accord with Schrödinger, albeit with a very simple example.

In this context, Watson and Crick's insight into the rules for base-pairing is central to the concepts of biological information and negentropy. The example of the one-bit electronic gate, which can exist in two distinguishable states, requires erasure and resetting in a single cycle. The formation of a single base-pair requires two bits – the minimum for any interaction involving two distinct choices – and is accompanied by a reduction in intrinsic entropy. Conversely its dissociation into its two constituents increases entropy. A base-pair represents an interaction dependent on information and thus satisfies both Chaisson's physical and operational definitions of complexity (Box 1.5). In this light it can be regarded as the simplest possible manifestation, or at least progenitor, of a complex state. In a living cell the number of interactions between macromolecules of all types is immense and requiring a correspondingly large quantity of information. Although some, maybe many, of these interactions are transient, they contribute to a dynamic ordered, or low entropy, state.

However a discussion of the potential information contained in a single base-pair is overly simplistic because in a sequence the information depends on context and composition or heterogeneity. A longer base sequence containing just a single type of base is information poor, while a sequence containing all four bases in equal proportions is potentially more information rich. Exactly the same principle applies to peptide sequences. For a three-base sequence, such as a codon, the computed Shannon entropy (bits/codon) in a longer random sequence is 3 (or 1/base) for a simple homopolymeric sequence, such as AAA, etc., but on average 6 for a sequence with 50% A-T and 50% G-C base-pairs (Nigatu et al., 2016). This reflects the number of combinatorial possibilities (Box 2.4). In an encoded mRNA the triplet code read sequentially three bases at a time the digital RNA sequence contains potentially three channels in different phases. Decoding the sequence and extracting the information content requires that the decoding apparatus identify the correct channel (usually only one) using appropriate start point information encoded in sequence. Identification of a start point by whatever mechanism ensures that the readout is selective and no longer random.

A physical consequence of sequence diversity can be envisaged by considering a population of a short homopolymeric polynucleotide, say polyA, mixed with complementary – in this case polyU – molecules of the same length. In such a mixture, the complementary polynucleotides will base-pair, but the number of possibilities is huge. They can form 1:1 complexes of the original defined length, but also each molecule can base-pair with several different molecules of its complement forming extended linear chains by overlapping interactions as well as branched networks. In this situation the probability of uniquely forming a 1:1 complex is extremely small. This probability can be increased somewhat by, for example, placing a unique four-base sequence (or its complement) in the centre of the original polynucleotides. By applying more stringent conditions to sequence alignment sequences, this heterogeneity reduces the number of structural possibilities. And so on. This is essentially a demonstration of Maxwell's aphorism 'the true logic for this world is the calculus of probabilities'.

This principle is not restricted to biological codes. A more common place example would be a monotonous sound. This could be split into successive repeats separated by a constant time interval. When the length of the repeats is varied the result is equivalent to a Morse code. Again heterogeneity is associated with an increase in the transmitted information. Or again, consider a crystal of common salt, or even a diamond. At the atomic level these are structurally well ordered but are essentially homogeneous, especially the latter. So probability of the location of a particular sodium ion or carbon atom to a defined position in the crystal is vanishingly small. This is in marked contrast to the probability of the occurrence of a particular amino acid at a defined location in a protein sequence or its structure. Because this is encoded, in the absence of any mutation the position is essentially invariant.

The biological implications of sequence heterogeneity are far-reaching. Within genomic DNA in the context of digital protein coding some sequences will be information rich and others information poor. Very generally coding sequences – those specifying protein sequences – are, by this definition, more information rich than non-coding sequences, especially including those between genes (Nigatu et al., 2016). But, non-coding regions, especially those containing transcriptional regulatory sequences can encode specific information presumably less encumbered by the constraints of digital protein-coding information. For example, the apical loops of supercoiled DNA plectonemes (Box 2.6) often contain several oligo(dA.dT) stretches, which bend the DNA (Laundon & Griffith, 1988). If coding for a protein this form of sequence organisation could restrict the range of amino acids in a translated protein.

Box 2.6 Plectonemes and Toroids

A plectoneme is a particular configuration of a DNA supercoil. In this high-energy form a circular DNA molecule adopts a structure in which the central region consists of two DNA duplexes wound around each other – the 'interwindings' – while the ends of the interwindings are closed by 'apical loops'. These loops and the interwindings are geometrically distinct. Another configuration adopted by supercoiled DNA is a 'toroid' in which the trajectory of the DNA resembles that of a spring. Toroids and plectonemes interconvert. Plectonemic DNA is found in bacterial chromosomes while toroidal packing is typically found eukaryotic chromosomes. The figure shows the forms adopted by negatively supercoiled DNA. For an exhaustive account of supercoil geometry, see Cozzarelli, Boles, and White (1990).

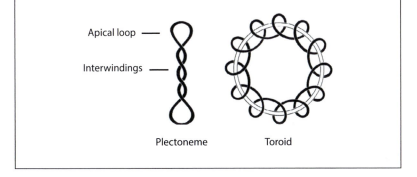

The proportion of coding to non-coding sequences varies both between biological taxa and also within taxa. This implies that differences in genome size are not necessarily reflected in comparable differences in potential information content. For proteins the situation is similar but because the genetic code is degenerate (a single amino acid can be encoded by several codons) sequence simplicity is proteins may not be reflected in a corresponding simplicity of the encoding nucleic acid sequence. Many proteins contain both simple sequence information-poor regions coupled to more complex information-rich regions. One example would be the chromatin associated linker histones involved in the compaction of chromatin fibres (Figure 2.6). These consist of a central sequence rich globular domain flanked by simple

N-terminal   Globular   C-terminal
domain   domain   domain

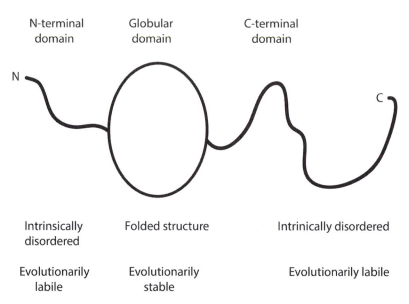

Intrinsically   Folded structure   Intrinically disordered
disordered

Evolutionarily   Evolutionarily   Evolutionarily labile
labile   stable

Figure 2.6 Domain structure of linker histones. N and C indicate N- and C-termini, respectively. The terms 'evolutionarily labile' and 'stable' are relative. The sequences of both the terminal and globular domains change over time but those of the terminal domains evolve more quickly.

sequences on either side. The globular domain folds into a defined structure while in solution the simple sequences remain intrinsically disordered and highly dynamic. Functionally the globular domain tethers the linker histone to a nucleosome in a defined position while the flanking regions, particularly the C-terminal tail, interact dynamically and with low specificity with the linker DNA between nucleosomes. Here high sequence complexity enables higher specificity in an informational interaction and is enabled chemically by a variety of distinct non-redundant contacts. Another characteristic of simple sequences is that they tend to exhibit more variation between related species. While their sequences are less conserved than in regions of high sequence complexity (Figure 2.6) the variation observed often retains the same (or a very similar) restricted set of amino acids with short tri- or tetra-amino acid motifs being shuffled. In the example of the C-terminal tail of histone this conserves the positive charge (from lysine and arginine residues) available to neutralise the negative charge on a particular length of linker DNA. Because there are many possible ways a histone tail can neutralise linker DNA this type of interaction is information-poor but is a necessary component enabling both chromatin dynamics and coacervation (Box 4.2).

Complexity as Information

A formal description of information as a process (Figure 2.5), as understood by Szilard, sheds little light on how information can act to increase the probability of an otherwise unlikely event. In the examples discussed – the ON/OFF switch, the signal/sensor interaction, the annealing of complementary nucleic acid sequences – the 'informative interaction' has thermodynamic consequences, manifested at least in part by a reduction in intrinsic entropy. But this in itself merely describes, but does not alter the probability of, the interaction.

An insight into this problem can be gained by comparing the entropy associated with one bit of information on Landauer's model ($\sim 10^{-23}$ JK^{-1}) with that associated with the probability of assembling an *E. coli* cell from its constituent matter in the equilibrium state (3.2×10^{-12} JK^{-1}). If the converse of Lewis' maxim that '... gain of entropy eventually is nothing more nor less than loss of information ...' were correct this implies that the information content of a single *E. coli* cell is equivalent to approaching half a terabit (where a terabit is an eighth of a terabyte). Similarly the amount of information associated with double helical *E. coli* genome of 4×10^5 base-pairs is approximately a tenth of a megabit – assuming that each base-pair contains 2 bits (but see preceding section). This, of course, does not include the encoded but not decoded, signal in the DNA sequence. The difference between the thermodynamic information in DNA and that in the complete organism is enormous. Such simplistic calculations come with caveats. Notably, Landauer's value assumed that resetting is done infinitely slowly, which is not the case in biological systems and therefore the entropy associated with a biological bit is likely significantly greater than Landauer's limiting value. Nevertheless, even allowing for such an effect there is still a large excess of information in the organised cell to account for.

So how can information selectively increase probabilities? With both Maxwell's thought experiment and the Lego constructor, the suitably equipped decoder – be it demon or small person – is an ill-defined black box that must accomplish more than one task. In different words, the probability of the final outcome is enhanced not simply by a single 'informative interaction' but by a number of such interactions being introduced into a more 'complex' system. This complex system could then, in principle, both integrate and amplify any initial signal. Rather than a separate phenomenon complexity is thus a simply a measure of the interactions between 'informative interactions' and represents a higher order manifestation of information (Figure 2.7). In reality complexity can be considered as a three-dimensional network of interactions between interactions (Figure 2.8). Within such networks integration may be effected both by direct molecular interactions and also by the generation of signals by, for example, an enzymatic reaction producing a diffusible molecule. Such integration – a form of communication – is a manifestation of local order that potentially enables the localised accumulation of higher concentrations of chemical reactants or signals, which in turn promote increased rates of formation of informative interactions. It is this local order that is crucial to the operation of a complex system and possibly nowhere more so than in the biological realm. In biology local order is both ubiquitous and essential. Protein assemblies are organised to carry out a defined sequence of chemical reactions often resulting in the synthesis of highly complex molecules, biomolecules are targeted to different parts of a cell, the leaves of plants are arranged to maximise the trapping of photons. Protein assemblies are, of course, modules and it is this very modularity (Box 2.2), coupled with the interactions between modules, that is characteristic of robust, heterogeneous complex systems (Simon, 1962).

Informative biological interactions, are by their very nature, essentially dynamic entities. That is, in general and in contrast to electronic switches, they have a finite lifetime consequent on association and dissociation. This property implies that the information content of a complex system, however defined, as a whole is also dynamic. In biological systems the flux of information content is especially evident in, for example, the changing nature of ecosystems or even in senescence, which may be accompanied by a gradual diminution in the totality of informative interactions.

Because information is physical, complexity, as a higher-order manifestation of information requires energy for its maintenance. The more energy that is available, the greater the complexity that can be attained.

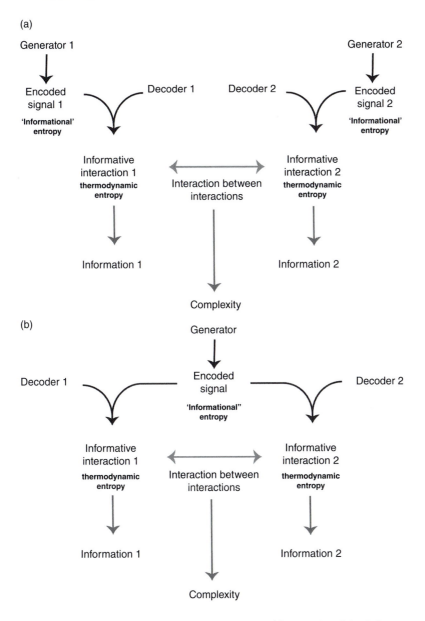

Figure 2.7 Complexity integrates informative interactions. (a) Integration of signals from different generators. (b) Integration of signal from a single generator selectively decoded by different decoders.

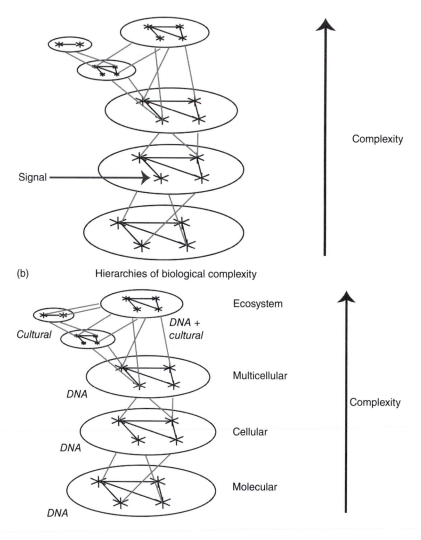

Figure 2.8 Hierarchical arrangement of complexity. (a) General principle. The upper levels (modules) are dependent on the lower levels. The connections within modules are more numerous than those between modules. Branching can generate diversity. (b) Complexity of biological organisation.

Regarded in this way the 3D complexity network can be represented as a hierarchy such that the higher levels, requiring more total energy, are dependent on the lower levels (Figure 2.8). Available energy would then determine the upward extent of the hierarchy. If sufficient energy is available, complexity has the potential to increase, if insufficient complexity would be reduced. Additional information from an external signal would have the potential to feed into the network at any level and so alter the overall pattern of interactions. Within a hierarchy branching may occur such that several higher levels of a comparable level of complexity can coexist but these levels only interact indirectly (Figure 2.8). Analogies in biology would be the phenomenon of adaptive radiation where a variety of different species evolve from a single ancestor and generate diversity with geographic separation. The varied giant tortoises on the different Galapagos Islands and the *Echium* (bugloss) explosion in the Canary Islands are two examples of many (Böhle, Hilger, & Martin, 1996). Again an ecological network dominated by a 'top predator' has a hierarchical structure.

The Nature of the 'Codescript'

The nature of the 'codescript' bears directly on the issue of the biological 'information' A DNA 'codescript' is a one-dimensional linear base sequence assembled on a chemical scaffold. Yet the 'instructions' it contains specify complex three-dimensional molecules capable of inter-acting directly both with themselves and with many others. This difference between the code and its ultimate product is fundamental to the functioning of biological systems and at its extreme essentially represents a transformation from the digital code in DNA to its structural translation in proteins.

In its simplest form the codescript is the succession of bases connected by a sugar-phosphate backbone within a single unfolded nucleic acid strand. In principle this strand can be either DNA or RNA. Yet the nature of the expressed information depends not only on the digital sequence but also on the scaffold on which it is constrained. Nucleic acid polymers can be single or double stranded, they can be circular or linear, they can fold into more complex structures stabilised by internal base-pairing interactions. Moreover the structures formed by both double and single nucleic acid strands are often labile and can change between different states. All of these different forms contain all or part of the codescript.

In contrast the translation of a protein coding sequence into a polypeptide chain produces a molecule where the sequence of amino acids specifies the mode of folding into a three dimensional structure. Some proteins fold into a very precise configuration, others are unstructured and for others folding is induced by contact with another molecule, frequently another polypeptide chain. Often in a single chain some sequences are highly structured and others not at all. In a protein the mode of folding depends on the sequence neighbourhood of particular amino acids. And two distant parts of a polypeptide chain can interact to stabilise a structure then the separation of the relevant amino acid sequences can also be important (Figure 2.9).

The formation of secondary structures by nucleic acid chains mixes both digital and environmental information. In double-stranded DNA the formation of base-pairs between a strand and its complement is entirely a digital interaction. Yet this conversion of the codescript vehicle from two single-stranded molecules to a single double stranded one introduces additional – emergent – chemical properties, including DNA bending and sensitivity to localised strand separation, which are highly dependent on the aggregate local characteristics of the DNA sequence – in other words 'analogue' properties (Box 2.7). The helical repeat of double-stranded DNA is likely the analogue property, which is centrally important to the mechanisms enabling biological information processing and which also likely determines the functional limits of biological complexity (see Chapter 9). It is defined as the number of base-pairs/double helical turn and is the reciprocal of the average intrinsic twist between successive base-pairs. Being a measure of the twist of the double helix the helical repeat changes the DNA duplex untwists to a single-stranded state. In solution it has an average value of ~10.5 bp/turn (Wang, 1979; Rhodes & Klug, 1980) but as the DNA untwists this value increases until it attains infinity on strand separation. Its biological importance rests on its role as an environmental sensor. It is increased at higher temperatures and by negative supercoiling but is decreased by higher ionic strengths, particularly increases in K^+ and divalent cations such as Mg^{2+} and Ca^{2+}. Many of the mechanisms regulating transcription can be understood as

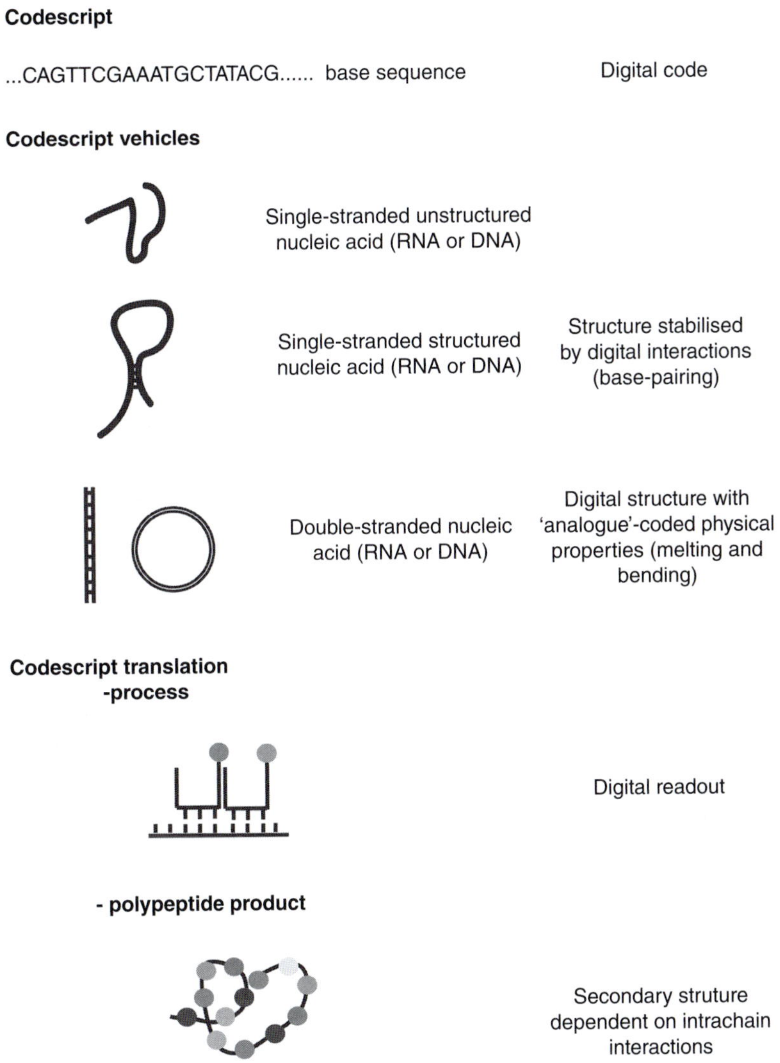

Codescript

...CAGTTCGAAATGCTATACG...... base sequence   Digital code

Codescript vehicles

Single-stranded unstructured
nucleic acid (RNA or DNA)

Single-stranded structured
nucleic acid (RNA or DNA)

Structure stabilised
by digital interactions
(base-pairing)

Double-stranded nucleic
acid (RNA or DNA)

Digital structure with
'analogue'-coded physical
properties (melting and
bending)

**Codescript translation
-process**

Digital readout

- polypeptide product

Secondary struture
dependent on intrachain
interactions

Figure 2.9 Different manifestations of the DNA codescript.

homeostatic responses to environmentally-induced changes in the
helical repeat (see Chapter 9).

Put another way, although double-stranded DNA is a vehicle for
maintaining the codescript and is dependent on digital interaction,

---

Box 2.7 Digital and Analogue Information

In the context of DNA the terms digital and analogue have been used respectively to describe the discrete nature of the sequence and the physical properties dependent on a succession of base-pairs. Strictly the latter remains a digital property because it represents the summation of adjacent individual digital components, i.e., it is not truly continuous. Only when the sequence is of infinite extent would the summation approximate to an analogue continuum. In the text the term analogue is used as defined. A useful approximation of analogue information is the averaged helical repeat of DNA, particularly the intrinsic repeat because this parameter takes into account a more extensive property of the DNA molecule. Because adjacent base-pairs interact chemically by stacking on each other the helical repeat is not independent of the digital base sequence because individual dinucleotides (base-steps) in the sequence adopt preferred and context-dependent twist values (Gorin, Zhurkin, & Olson, 1995; El Hassan & Calladine, 1997; Olson, Gorin, Lu, Hock, & Zhurkin, 1998).

---

some of its important, functional properties are principally 'analogue.' It is an information repository but cannot in itself act a direct coding device for the synthesis of a protein. It is more equivalent to hard disk storage than to an active decoding role. The local folding of single stranded RNA molecules can be regarded as driven by essentially digital information. The localised base-pairing of separated sequences to form a double stranded region with an intervening loop is digital in nature yet the size of the loop – and hence the final structure – depends on the number of base-pairs between the hybridising sequences (Figure 2.4) – again digital. In this example, as also in proteins, the resulting structure depends on sequence context.

Thus thanks to the codescript, nucleic acids – in their varied forms – utilise digital information as a guide for structure. In such a scheme base-pairing is a fundamental attribute – one that is apparent not only in DNA–DNA and RNA–RNA interactions but also in heterologous RNA–DNA pairing. The gamut of possible base-pairing combinations provides the functional basis for replication, transcription, translation, recombination and for much of the regulation of both transcription and translation.

Ribonuclease P: A 'Simple' Example of Biological Complexity

One example of the manifestation of complexity as information is the relatively simple RNase P enzymatic system. This bacterial enzyme – which can be regarded as a mini-module – is made up of two components, an RNA molecule with enzymatic activity (it is a ribozyme) and a short polypeptide. The function of the ribozyme component is to cleave a leader sequence from a tRNA precursor to generate a tRNA (Figure 2.10). Cleavage by the ribozyme alone is relatively inefficient yet when bound to the polypeptide the rate of cleavage is enhanced ~200 fold. The polypeptide itself possesses no enzymatic activity but by binding close to the active site facilitates the enzymatic process. Here a

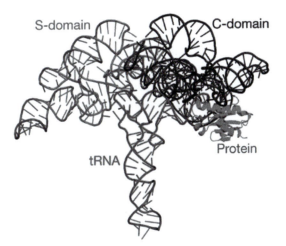

Figure 2.10 A representation of the crystal structure of ribonuclease P complexed with a tRNA molecule (red). The enzyme comprises an RNA component (blue) and a protein component (green). The active centre catalysing the cleavage of the tRNA precursor is located in the dark blue cavity adjacent to the protein. *A black-and-white version of this figure will appear in some formats. For the colour version, refer to the plate section.* Source: Reproduced with permission from Reiter et al. (2010). Copyright 2010, Springer Nature

two-component system (RNA plus polypeptide) is far more effective
that a single component.

To achieve both the capability of the ribozyme alone and of the
complex requires 'information'. The folding of the RNA molecule into
a relatively defined tertiary structure is buttressed by 18 base-paired
double-stranded regions, while the polypeptide is again folded into a
distinct tertiary structure stabilised by internal interactions. Not only
does the structure depend on intramolecular contacts but also on
specific intermolecular contacts between the ribozyme RNA and the
polypeptide and also between the ribozyme RNA and its pre-tRNA
substrate. (And of course the tRNA itself contains several double-helical
regions.) Thus the number of 'informative interactions' necessary for
the efficient functioning of this simple complex is correspondingly large
($\sim 10^3$). These interactions are, of course, encoded in the sequence signal,
but nevertheless they represent a substantial amplification of the
thermodynamic negentropy in the corresponding double-helical DNA.
Although this single example is only illustrative, it is easy to see how
such a change, especially with the multitude of examples in a cell,
reflects the vanishingly small probability calculated by Morowitz for
the assembly of a bacterium arising from a spontaneous fluctuation in
an equilibrium system.

Information Storage

The capacity of any system for information storage is normally
limited, either by simple physical constraints and/or by the need to
integrate different types of information. Biological information is no
exception. Here the limitations depend on the physical and chemical
properties of the encoding molecules. But this is not the only constraint. The evolution of information storage is also critical because an
established mechanism, even if not potentially the most efficient, can
exclude the expansion of coding capability. A simple example is the
genetic code. Once the operation of a non-overlapping triplet code was
established the number of triplets, assuming the use of all four bases, is
immediately limited to 64. But, because a full triplet code was likely
preceded by a more primitive interrupted doublet nature (Chapter 3)
the number of possible combinations encoding an amino acid would,

on Bayesian principles (Box 1.6), be substantially lower. In this situation once all the triplets have been assigned to an amino acid, including the no-amino acid stop signal, further expansion of the code utilising a simple triplet code would be largely precluded. Instead expansion would rely on the adoption of different modes of sequence recognition of the mRNA.

Informational Hierarchies

The genetic code well illustrates the mechanisms of information processing in biological systems. The translation of an mRNA molecule requires the operation of a decoder (tRNA plus ribosomes), which is itself encoded in the genome. The result is to transform the selected copy of the codescript in RNA into a polypeptide chain, which can participate in both internal (determining its structure) and external informative interactions. By itself an mRNA molecule would most often participate in informative interactions with a complementary nucleic acid sequence – most frequently a tRNA molecule. Essentially the decoder here acts as an amplifier greatly increasing the amount of expressed 'information'. The same principle operates when a DNA-binding protein molecule binds in a sequence-specific manner to the DNA duplex (or indeed to an mRNA molecule). In this example the sequence is the signal and the DNA-binding protein the decoder. Again the decoder is itself encoded in codescript.

mRNA translation and protein-nucleic acid recognition are only two examples of several modes of information transmission. Another is the epigenetic tagging of both DNA (by, for example, DNA methylases) and of nucleosomes by a suite of protein modifying enzymes (Box 2.8). Such tagging can modulate specific recognition by regulatory proteins enabling discrimination between tagged and unmodified sites. Yet again the role of DNA in this scenario is to provide the logical code – in this case a signal – for information processing (Figure 2.11). Changing the DNA sequence has the potential to change site tagging. DNA thus remains the primary source of information. Information processing at the levels of, say, epigenetics or even the translation of mRNA could be as sub-routines in the overall DNA-specified programme. The sub-routines ultimately depend on the DNA sequence.

Box 2.8 Epigenetics

As for both information and complexity the definition of epigenetics has varied but in its more exclusive form is taken to be the study of heritable phenotype changes that do not involve changes to the DNA sequence (see Chapter 4 for further discussion). A more inclusive definition is 'the structural adaptation of chromosomal regions so as to register, signal or perpetuate altered activity states' (Bird, 2007). Thus more generally epigenetic mechanisms maintain a genetic programme or process in the absence of mutation. These changes, which are often ephemeral, can result from chemical modifications of chromatin structure, which can include DNA methylation, histone modification, and RNA processing. The modern genesis of this concept arose from the pioneering studies of Barbara McClintock (1950) on maize. She showed that a genetic locus on one chromosome could move to another and in so doing changed the colour of the maize kernels, thus introducing the notion of 'jumping genes' (see Chapter 6 for further discussion). Strictly such a translocation changes the DNA sequence at the initial and final sites, although because jumping is often highly precise overall no base-pairs are lost or gained. However transposition is correlated with DNA methylation, usually at the 5 position of cytosine. In animals this modification marks specific DNA sequences in the gametes from different parents – genomic imprinting – and is a determinant of expression from male and female alleles. Many histone modifications are associated with changes in the regulated expression of genes and often determine the development of a cell lineage during development.

In his original discussion Schrödinger emphasised the thermodynamic implications of increasing biological organisation. The question is then how the codescript – the information – contributes to 'negentropy'. A possible answer can be visualised through the prism of structure and function. The primitive codescript, RNA, is a digital code with 4 characters. However, the genetic code, in the form of encoding mRNA, while still a digital code, is both a triplet code and phased, i.e., it contains both start and stop signals but contains 20 characters. As digital codes both a nucleic acid sequence and an amino acid sequence can be considered as Shannon information (see Muskhelishvili [2015]

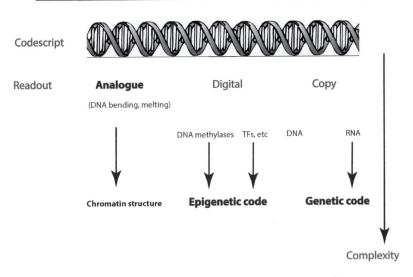

Figure 2.11 DNA specifies sub-routines for informational processing.

for an extensive discussion). By an emergent stratagem the 4-character RNA code became the 20-character amino acid sequence in proteins. In principle doublet and triplet genetic codes could encode 16 and 64 amino acids respectively but partly because the code has a Bayesian character the actual present day numbers are considerably lower (see Chapter 4 for discussion).

Both the 4- and 20-character codes specify structures but because the number of combinatorial possibilities is very much greater for a 20-character code, the structures so specified can be more precisely delineated – they are less fuzzy. In a long RNA molecule the 4-character code often creates an assemblage of alternatively base-paired structures – a phenomenon that is utilised in the cellular adaptation to low-temperatures. For proteins, which can be considered as the ultimate expression of the codescript, the ability to form a relatively more stable folded structure enables the enzymatic catalysis of a large range of chemical reactions and a consequent increase in the efficiency of energy conversion to create complex organic molecules. In contrast the current range of reactions catalysed by ribozymes is relatively small being essentially limited to RNA processing and peptide bond formation. (Chemically the actual possible range may be larger but its further evolution would likely have been stymied by the appearance of protein-catalysed reactions.)

Both folded RNA structures and stable folded proteins represent the thermodynamic manifestations of the codescript, and the difference is a consequence of a step-change in information processing dependent on an evolutionary innovation in the decoding device. The amino acid code is more complex that the simple RNA digital code and this greater complexity can be reflected in greater structural precision and consequent organisation. In other words the probability of a particular amino acid occupying a particular three-dimensional location in a folded protein is on average higher than that of the probability of a particular base occupying a defined position in an RNA structure. In essence biology is a probability game. In informational terms Landauer's description of the interaction of a signal with a decoder has a thermodynamic consequence as also does, say, the interaction of a photon with a biological receptor – essentially a decoder. The resultant organisation then depends on the complexity of the signals, i.e., how they are encoded, and/or on their integration. At the risk of incurring Jaynes' posthumous wrath it seems reasonable to conclude that there is a clear link between the encoded Shannon signal in Schrödinger's codescript in a DNA or RNA sequence and the thermodynamic organisation manifest in biological complexity.

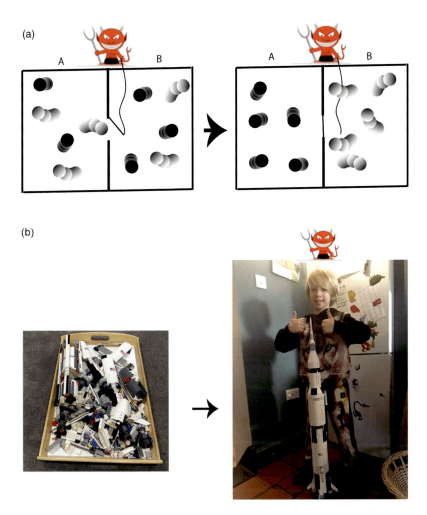

The Force (Maxwellian) Awakens

Figure 2.1 (a) Maxwell's thought experiment. The demon 'knows' which gas molecules are hot (grey) or cold (black) and by allowing only cold molecules to move to the left through the door and hot molecules to move to the right effects a separation into two homogeneous compartments, A and B. Source: Figure redrawn from Wikipedia.org and reproduced under the terms of the Creative Commons Attribution-Share Alike 3.0 Unported licence. (b) The importance of instructions. A pile of Lego bricks transformed into a model of a rocket by a small boy (the author's grandson). Maxwell's demon approves. Source: Photos by H. Travers and J. Burt

Visible UV

Figure 2.3 A hawkweed flower visualised with light of different wavelength, as perceived by a human (left) and a bee (right). Source: Reproduced from Wikipedia.org under the terms of the Creative Commons Attribution-Share Alike 3.0 Unported licence

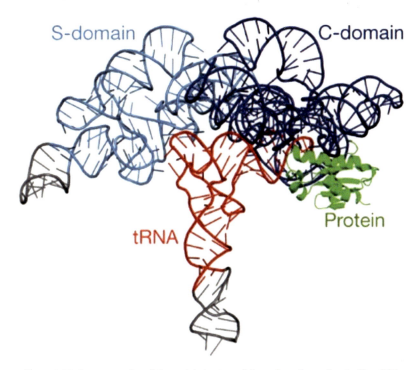

Figure 2.10 A representation of the crystal structure of ribonuclease P complexed with a tRNA molecule (red). The enzyme comprises an RNA component (blue) and a protein component (green). The active centre catalysing the cleavage of the tRNA precursor is located in the dark blue cavity adjacent to the protein. Source: Reproduced with permission from Reiter et al. (2010). Copyright 2010, Springer Nature

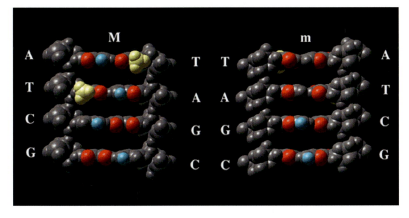

Figure 3.2 Exposure of chemical groups of nucleotide bases in the major and minor grooves of DNA. M, major groove; m, minor groove. Yellow, thymine 5-methyl group; blue, basic groups: adenine 6-amino group (major groove), cytosine 4-amino group (major groove) and guanine 2-amino group (minor groove); red, exposed cyclic nitrogen atoms and oxy-groups. Note that the presence of the thymine 5-methyl group in place of hydrogen in uracil the enables A-T base-pairs to be distinguished from T-A base-pairs in the major groove. Source: Adapted with permission from IMB Jena Image Library

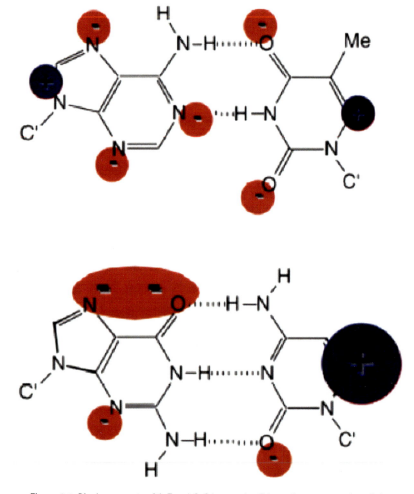

Figure 3.4 Dipole moments of A-T and G-C base-pairs. Schematic representation of the electrostatic potential at the van der Waals surface of the DNA base-pairs. Regions of positive (blue) and negative (red) charge density are marked. (top) A-T; (bottom) G-C. In G-C base-pairs but not A-T base-pairs the potential distribution is highly asymmetric, creating a dipole. Source: Reproduced with permission from Hunter (1996), *Bioessays*, *18*, 157–162

Figure 3.11 The Japanese 'canopy plant', *Paris japonica*, contains the largest recorded genome size in angiosperms. Source: Photograph in the public domain, reproduced from Wikimedia

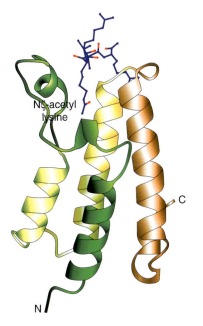

Figure 4.4 The bromodomain of the yeast histone acetyltransferase Gcn5 binding to its ligand N-acetyl lysine. Source: Reproduced with permission from Owen et al. (2000), with permission from John Wiley & Sons, Ltd. Copyright 2000 European Molecular Biology Organization

Figure 5.1 Complexity from genome sharing. A lichen (*Caloplaca saxicola*) growing on a stone fence post. Source: Photograph by the author

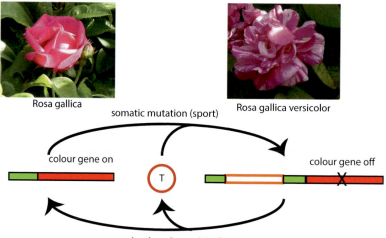

Rosa gallica

somatic mutation (sport)

Rosa gallica versicolor

colour gene on

colour gene off

T

reversion (precise excision)

Figure 6.2 Variegated patterns in plant flowers and leaves can be due to the insertion of a short DNA transposon (T) in the control region (green) of a gene. When this DNA jumps out (precise excision), the original DNA organisation and the phenotype is restored. The different variants of *Rosa gallica* shown arose from the same rootstock in the author's garden.

Figure 7.1 Darwin's warm little pond? A Kamchatka mud-pot. Source: Reproduced from Dibrova, Galperin, Koonin, & Mulkidjanian (2015) with kind permission of Anna Karyagina

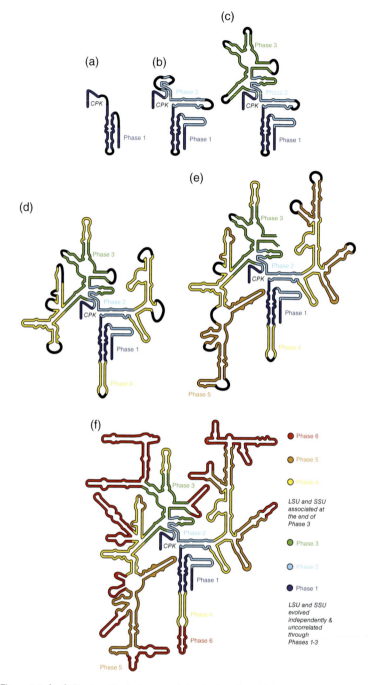

Figure 9.1 (a-e) The hypothesised sequential evolution of the RNA in the small ribosomal subunit from simple short RNA molecules existing in the primeval RNA world to the present-day association of much longer RNA molecules with proteins. The evolutionary stages are colour-coded to correspond with successive sequence insertions. Source: Adapted from Petrov et al. (2015) Proceedings of the National Academy of Science, USA, 112, 15396–15401.

3

DNA

The Molecule

L'invariant biologique fondamentel est l'ADN.
—*Jacques Monod, 1970*

Today the feature of DNA that defines the molecule is the fact that the two strands are entwined as a right-handed double helix. In common parlance, DNA is 'the double helix'. While this double-helical character is not required by the base complementarity per se – a simple straight ladder structure would fulfill this function just as well – it does impart crucial physical and chemical properties to the polymer. It is these properties that play a major role in the biological function of DNA. The genetic functions of DNA can thus be understood as the synergism of two properties – a tape containing the information store encoding the sequences of proteins and RNA molecules and a polymer existing as double-helical string enabling the packaging, accessibility and replication of the information store. Crucially not only the coding of proteins and RNA molecules but also the physicochemical properties of the polymer are specified by the base sequence.

The Canonical Base-Pairs

A DNA strand is essentially a string containing the four canonical bases – adenine (A), cytosine (C), guanine (G) and thymine (T) – ordered in an apparently irregular array. One of the earliest clues to the nature of the DNA code was the observation by Erwin Chargaff and his colleagues that in the purest DNA preparations the molar ratio of the purine bases (A + G) to pyrimidines (C + T) was very close to 1 (Zamenhof, Shettles, & Chargaff, 1950). Also particularly striking was the equivalence of A to T and of G to C. It was this equivalence that led Watson and Crick to the concepts of the specificity of base-pairing and the nature of the complementary base sequences in the two strands of the double helix (Figure 3.1).

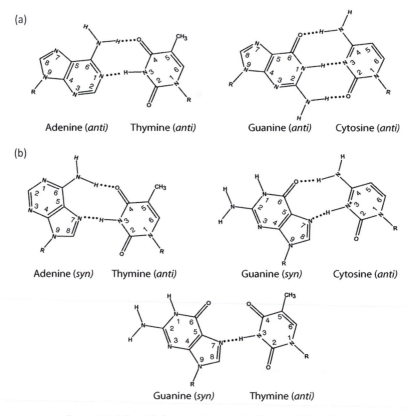

Figure 3.1 A-T and G-C base-pairs shown for Watson–Crick pairing (a) and Hoogsteen pairing (b). Structures of alternative base-pairs maintaining Watson–Crick hydrogen bonding; *anti* and *syn* refer to alternative conformations of the attached deoxyribose residue. Source: Reproduced from Johnson, Prakash, and Prakash (2005), *Proceedings of the National Academy of Sciences USA, 102*(30), 10466–10471, with permission

Not only did the double-helical structure provide a crucial insight into the mechanism of genetic inheritance, but it also pointed to the operational bounds of the biological system that it encodes. In existing cells most processes of information transfer involving DNA – DNA replication, the copying of selected sequences of DNA into RNA and genetic recombination – require the melting of a limited region of the double helix to form a stretch of DNA containing two separated single strands – albeit connected to contiguous double-stranded regions. Strand separation occurs sharply over a narrow range of temperature. If the temperature is too high, the whole DNA molecule is converted to single strands and the integrity of genetic material could be lost or at least severely compromised. If it is too low, the energetic barrier to efficient information transfer is substantially greater. Small wonder then that organisms have evolved multiple mechanisms for fine-tuning the melting temperature of a DNA sequence to be compatible with their immediate environment.

DNA melting is a consequence of the unwinding of the double helix. It is not, as so often depicted in textbooks, simply a pulling apart of the two strands to form a 'bubble', but a decrease in the DNA twist such that the number of base-pairs per turn increases from the average 10.5 to infinity. To mitigate any effects of more extreme temperatures on DNA function, organisms modulate DNA twist. For example, in certain hyperthermophiles, temperature-dependent decreases in DNA twist can be counteracted by the enzyme reverse gyrase. Conversely, in bacteria a cold shock is accompanied by an increase in the expression of an enzyme, DNA gyrase, that has the potential to decrease DNA twist. Again, DNA twist is also influenced by the ionic composition of the intracellular milieu, and changes in this parameter can parallel changes in DNA morphology during both a bacterial growth cycle and the formation of metaphase chromosomes during mitosis. Ultimately, however, DNA-dependent information transfer is restricted to the temperature range that currently dominates the biological world.

Energetically the coexistence of two types of base-pairs in the DNA helix has important consequences. Because base-pairing for G and C and for A and T depends, respectively, on three and two hydrogen bonds, A-T base-pairs melt at a lower temperature than G-C base-pairs (Figure 3.1). This means that transient unwinding of the double helix occurs more frequently in A-T–rich regions. But similar considerations also apply to RNA because it too can adopt a double-helical configuration with the same base-pairing rules, and consequently the stability of codon–anticodon interactions and of hairpins in otherwise single-stranded RNA molecules may depend on base-pair composition.

Alternative Modes of Sequence Recognition

The gold standard of sequence recognition in the double helix employed in both transcription and replication utilises the base-pairing rules formulated by Watson and Crick. However, as they acknowledged, bases can pair in different ways. In particular, a different base-pairing geometry, the Hoogsteen base-pairs, in which the purine base is rotated relative to that in the standard base-pair, reduces the distance between the C1 carbon atoms of the associated sugar moieties (Figure 3.1). Although this type of base-pairing is incompatible with the structure of the canonical DNA double helix, it is found in other structural forms of DNA, notably H-DNA and G-quadruplexes (see following). Similarly yet another type of non-canonical base-pair has been postulated to stabilise the i-motif formed by sequences complementary to those forming G-quadruplexes.

Another form of DNA–DNA interaction that has received relatively scant attention is the ability of two double helices of the same sequence to align with each other (Inoue, Sugiyama, Travers & Ohyama, 2007; Baldwin et al., 2008). The attractive force causing this DNA self-assembly could function in biological processes such as folding of repetitive DNA, recombination between homologous sequences and synapsis in meiosis. But how is this type of homologous recognition effected? A possible mechanism is the mutual alignment of the electrostatic signature of a base sequence. Another, not exclusive, suggestion is that the flipping out of bases from the double helices may also be involved. This mode of sequence recognition likely lacks the precision provided by the Watson–Crick base-pairing rules but nevertheless should facilitate initial discrimination in searches for similar or identical sequences.

DNA Information

What is the nature of the genetic information stored in DNA? The distinction between a linear code responsible for specifying the sequences of RNA and protein molecules and also sequence-specific recognition by DNA-binding proteins, and an equally important more continuous structural code, specifying the configuration and dynamics of the polymer, extends the informational repertoire of the molecule. Both these two DNA information types are intrinsically coupled in the primary sequence organisation, but whereas the linear code is, to a first approximation, a direct digital readout, the structural code is determined not by individual base-pairs, but by the additive interactions of successive base steps. The latter code, being locally more continuous, thus has an analogue form.

Direct and indirect readout of DNA-recognition sites by proteins is a major determinant of binding selectivity. In direct readout the individual bases in a binding sequence make direct and specific contacts to the protein surface whereas in indirect readout the binding affinity depends on recognition of a structure, such as a DNA bend or bubble, whose formation is influenced by DNA sequence, but does not in general require a protein contacting a specific base. In practice DNA recognition by proteins effectively spans a continuum from completely digital to completely analogue with many proteins utilising both modes.

For both modes of recognition the DNA double helix differs from, and is arguably more effective than, the RNA double helix. Direct readout requires intimate contact between exposed chemical groups on both the protein and nucleic acid surfaces. For DNA recognition direct readout in most examples takes the form of a DNA-binding protein fold being inserted into the major groove. In this groove different charged groups of the bases in a pair are exposed to those in the minor groove (Figure 3.2). Consequently while A-T and T-A base-pairs in a sequence are distinguishable by charge pattern in the major groove, in the minor groove they are not. The major groove thus provides more sequence information than the minor groove. However, importantly, the wide and shallow morphology of the DNA major groove is in stark contrast to the narrow and deep structure of the RNA major groove. This pattern is reversed for the minor groove.

For a protein DNA-binding motif, particularly one containing an α-helix, access to the DNA major groove is more facile than to the minor groove. This fundamental difference between DNA and RNA follows directly from their chemical structures. Whereas DNA can adopt (at least) two forms of right-handed double-helical structures, A-DNA and

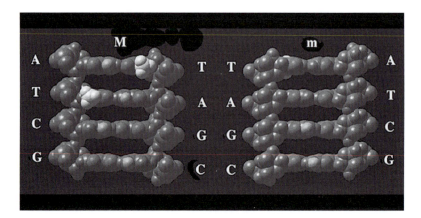

Figure 3.2 Exposure of chemical groups of nucleotide bases in the major and minor grooves of DNA. M, major groove; m, minor groove. Yellow, thymine 5-methyl group; blue, basic groups: adenine 6-amino group (major groove), cytosine 4-amino group (major groove) and guanine 2-amino group (minor groove); red, exposed cyclic nitrogen atoms and oxy-groups. Note that the presence of the thymine 5-methyl group in place of hydrogen in uracil the enables A-T base-pairs to be distinguished from T-A base-pairs in the major groove. *A black-and-white version of this figure will appear in some formats. For the colour version, refer to the plate section.* Source: Adapted with permission from IMB Jena Image Library

B-DNA, RNA can only form an A-type double helix because of the steric restrictions imposed by the $2'$ hydroxyl residue on ribose (Figure 3.3). The B-DNA structure, that proposed by Watson and Crick, is most stable at high humidity but converts to the A-form as the water activity is lowered. On this argument it is the ability to form the B-form that facilitates direct access to DNA sequence information. Not only does the A -> B transition affect direct readout, but it also changes the physicochemical properties of the polymer. An A-type double helix is, on average, stiffer than a B-type double helix, and consequently distortion of A-DNA to a particular bent configuration is energetically less favourable than for the corresponding distortion in B-DNA. Such differences would be expected to favour B-DNA as the preferred substrate for packaging involving tight DNA bending.

Although the formation of a B-type structure is a crucial aspect of DNA functionality, the factors that shift the A <-> B equilibrium are, apart from water activity, poorly understood. One aspect is base-type. In principle the coding capacity of DNA can be achieved not only by the canonical A-T and G-C base-pairs but also by other possibilities.

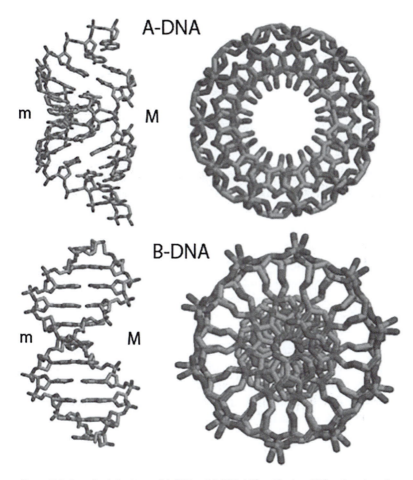

Figure 3.3 Formalised structures of A-DNA and B-DNA. LHS – side view; RHS – view along the helical axis. In B-DNA the major groove (M) is wider than that in A-DNA, and so base-specific information is more accessible in B-DNA. Note that in B-DNA the base-pairs are centrally aligned across the helical axis whereas in A-DNA they are more peripheral. Source: Adapted from Arnott (2006), *Trends in Biochemical Sciences, 31*, 349–354 with permission

For example, a DNA polymer with diaminopurine-thymine (DAP-T) and hypoxanthine-cytosine (H-C) base-pairs with, respectively, three and two inter-base hydrogen bonds would, in principle, present a similar potential for protein recognition. Other variations would be DNA molecules in which all the base-pairs contain either two or three hydrogen bonds. However not only do the component bases specify a digital code, but they also affect the physicochemical properties of the molecule. For example, DNA molecules with a reversed pattern of hydrogen bonding (DAP-T and H-C base-pairs) more readily adopt an A-type conformation than DNA with the canonical base-pairs. This is because the properties of the double-helix depend not only on the base-pairing capacity of the constituent bases but also on the chemical interactions – the stacking interactions – between adjacent base-pairs in a sequence that determines the chemical stability of an individual base-step (comprising two successive base-pairs) and cumulatively that of a whole DNA molecule. In a double-helical DNA molecule the ability of base-pairs to stack on top of each other depends on two factors – the geometry of the stacking base-pairs and also the charge distribution along the base-pairs. The geometry of base-pairs is little changed by altering the chemical groups that are exposed in the grooves because such changes do not alter whether a particular base is a purine or a pyrimidine. However, adding or removing such a group can profoundly affect the charge distribution across a base-pair and in some cases concomitantly change the base-pairing interactions. A good example is provided by the 2-amino group of purines, which is exposed in the minor groove (Figure 3.2). The removal of the 2-amino group from guanine creating hypoxanthine and its addition to adenine creating diaminopurine effectively switches the number of hydrogen bonds associated with G-C and A-T base-pairs and does not, in principle alter the coding capacity of DNA. However, the charged 2-amino group, by being in a different immediate chemical environment, also affects the charge distribution or dipole moment associated with individual base-pairs and consequently directly affects the ability of base-pairs to stack on top of each other in the double helix (Figure 3.4). But why should changes in stacking between base-pairs affect the transition between the A- and B-forms of DNA? Part of the answer probably lies in the disposition of base-pairs across the helical axis of DNA. When A-T base-pairs stack on top of each other, as they would in base-steps such as ApT and also ApA with its complement TpT, the base-pairs are aligned along the DNA so that the base-pair is more or less centred on the helical axis By contrast, when G-C base-pairs stack on top of each other, as in the base-pair GpG with its complement CpC, the base-pairs are no longer centred on the helical axis but are instead displaced slightly towards the periphery of the double-helix. One of the defining

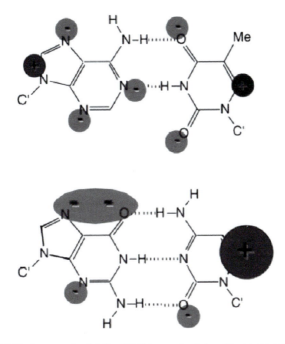

Figure 3.4 Dipole moments of A-T and G-C base-pairs. Schematic representation of the electrostatic potential at the van der Waals surface of the DNA base-pairs. Regions of positive (blue) and negative (red) charge density are marked. (top) A-T; (bottom) G-C. In G-C base-pairs but not A-T base-pairs the potential distribution is highly asymmetric, creating a dipole. *A black-and-white version of this figure will appear in some formats. For the colour version, refer to the plate section.* Source: Reproduced with permission from Hunter (1996), *Bioessays, 18*, 157–162

differences between A-DNA and B-DNA is the position of the base-pairs relative to the helical axis. Whereas in B-DNA they are largely centred on the helical axis, in A-DNA they are significantly displaced so that the base-pairs themselves form a spiral. This is not the only significant structural difference between A- and B-DNA – the conformation of the sugar-phosphate backbones also differ in important respects. The transition point between the two forms is likely to depend on several factors, one of which would be the energetically preferred average position of the base-pairs. Changing the latter by base-substitutions would then change the transition point. In other words, the ability to assume the B-conformation, which confers on DNA an important aspect of its unique genetic role, would itself be dependent on base-type. The initial question, 'Why DNA?' can then be refined to 'Why A-T and G-C base-pairs and not other combinations with the same coding capacity?' It is possible, but not yet proven, that the current overwhelming preference for A-T and G-C base-pairs in DNA is dictated by the energetics of base-pair stacking relative to other combinations. While this might constitute a reason for the selection of these base-pairs in DNA, no such simple argument can be advanced for their, presumably, prior use in RNA, although even in RNA the stability of different base-steps and hence of the double helix itself is also likely dependent on the precise nature of the constituent base-pairs.

DNA as a Conformationally Flexible and Dynamic Polymer

In genomes DNA molecules are generally very long, thin polymers with a diameter of 2 nm and a length that can extend to 10^8–10^9 nm. As an information store not only must DNA be able to encode the genetic information required to specify proteins, but also it should be packaged in a compact form that allows the accessibility of that information to be regulated. In turn the functional accessing of information may also involve structural changes in the double helix itself. However, the very nature of DNA – again an immensely long, very thin polymer – requires that within the cell the molecule be compacted into a small volume while maintaining accessibility. These requirements for compaction, accessibility and structural modulation imply that DNA be both flexible and able to change conformation in response to enzymatic manipulation.

The DNA molecule may be modelled as an extremely long, thin string of moderate elasticity that can be bent into the configurations required for packaging. Both the preferred direction of bending and the stiffness are sequence dependent. A directional bending preference, or bending anisotropy, facilitates the wrapping of DNA on a complementary protein surface. However, such a preference also implies that anisotropy increases the overall stiffness because it reduces the degrees of bending freedom. In other words, by reducing bending freedom, the intrinsic bending entropy of the sequence is reduced, and on binding to a preferred protein surface there is a corresponding reduction in the entropic penalty. Such bending preferences are important determinants of the binding of both enzymatic manipulators of DNA and the abundant so-called architectural proteins that direct the local packaging of the polymer. The archetypical example of this mode of packaging is the nucleosome core particle – the fundamental unit of DNA packaging in eukaryotic chromosomes – in which 145 bp of DNA are wrapped in 1.6 turns tightly around a histone octamer (Luger, Mäder, Richmond, Sargent, & Richmond, 1997). The signature of bending directionality is the presence of alternating short stretches of G/C-rich and A/T-rich DNA sequences in helical phase. Such an organisation confers bending anisotropy because G/C- and A/T-rich sequences favour, respectively, wide and narrow minor grooves. Consequently because of tightly bent DNA, both DNA grooves are narrowed on the inside of a bend and widened on the outside, G/C-rich sequences are favoured where the minor groove points outward and A/T-rich sequences where the minor groove points inward.

Not only must DNA be bendable in order to be packaged efficiently but also the copying of the DNA sequence during transcription or DNA replication requires the separation of the two strands of the double helix, a transition in which the double helix is untwisted to form a bubble. The initiation of copying at the points at which strand separation is nucleated is facilitated by highly localised less-stable

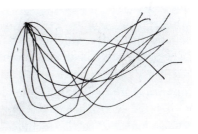

More flexible - isotropic   More rigid – anisotropic
                                   More sequence 'information'

Figure 3.5 Isotropic and anisotropic bending in DNA molecules. Some DNA molecules – isotropic – can bend freely in any direction while others – anisotropic – are stiff and bend in a preferred direction yielding a population of molecules with similar trajectories. Source: Reproduced from Georgi Muskhelishvili, with permission from 2016, Springer Nature

DNA sequences with lower stacking energies. Of these the least thermally stable base-step is TpA. Such localised untwisting of DNA affects not only DNA melting but also DNA bending. When a bubble is formed in the DNA double helix not only is its bending flexibility is increased, but also the directionality of sequence-directed bending becomes attenuated and more isotropic (Figure 3.5). In other words, under these conditions a DNA molecule no longer has a strongly preferred direction of bending.

DNA as an Energy Store

An often-overlooked function of DNA in a cell nucleus or bacterial nucleoid is that it can act as an energy store. This property is a direct consequence of the double-helical character of the molecule. Not only does it exist as a simple intramolecular interwound coil, but also the DNA chain can, under torsional stress, adopt a coiled configuration, or supercoil. Such supercoils have a higher intrinsic energy than DNA molecules not subject to torsional stress, aka 'relaxed'. Within both the eukaryotic nucleus and the bacterial nucleoid supercoiling is ubiquitous.

An open circle represents a state where the DNA molecule, under the prevailing environmental conditions, occupies an energetic minimum. It is 'relaxed'. However, enzymatic manipulation of DNA using energy from ATP can alter the torsional state of DNA, inducing more coiling – 'supercoiling' – within such a closed system. This coiling can be 'positive' – overwinding – in the same sense as the DNA double helix or 'negative' – underwinding – in the opposite sense. Because DNA is a right-handed double helix, an overwound DNA molecule contains more right-handed double-helical coils than the corresponding relaxed DNA molecule. Conversely, an underwound DNA molecule contains fewer right-handed coils than the same molecule in the relaxed state. This underwinding enables the facile separation of DNA strands at particular locations in the double helix.

Changing the number of coils in the double helix itself, equivalent to changing the number of base-pairs/double-helical turn, represents a change in the twist of the double helix (Box 3.1). An underwound DNA has more base-pairs/turn and an overwound helix fewer. Equally important, supercoiling can induce a change in the trajectory

---

Box 3.1 The Consequences of Being a Double-Stranded Helix

Two of the key functions of DNA are to serve as a template for both the production of single-stranded mRNA and for the replication of the molecule. Both these processes require the formation of a DNA bubble – a stretch of one to one and a half turns of DNA in which the two strands are melted and so completely separated. The mechanical consequence of the double helical structure is that the two strands cannot simply be pulled apart without affecting the structure of the rest of the molecule. In this context DNA melting necessarily involves a change in the average twist of the molecule and hence has rotational consequences.

Any rotation created by the formation of a DNA bubble will likely be transmitted along the DNA molecule. If the ends of the molecule are free to rotate, the generated rotation will be dissipated, but if not the DNA will supercoil, i.e., will adopt a coiled trajectory that can be either plectonemic or toroidal (see Box 2.6). This property directly follows from their double-helical nature of DNA.

---

of the double-helical axis – writhe at a lower energetic cost. This writhing can be manifested in two main forms as a toroid where the path of the DNA resembles a spring or as a plectoneme where the double helix wraps around itself forming a structure that resembles, for the most part, a length of twine or string (Figure 3.6). However, because the DNA double helix is continuous, the strands of the interwindings, where the DNA wraps around itself in a coil, are connected by tight loops of DNA at the ends of the molecule.

In bacteria the chromosomal DNA is often intrinsically supercoiled, and the free DNA is extensively writhed adopting a predominantly plectonemic form. In contrast, in the eukaryotic nucleus the super-helicity is differently distributed with the packaged DNA being constrained in a toroidal form (see Figure 2.6) of a left-handed (counterclockwise, and hence negative) wrapping of the DNA around an octamer of basic histone proteins

Enzymatic manipulation of DNA supercoiling can take two forms. On the one hand, an enzyme, termed a 'topoisomerase', can bind to a single site and directly change the coiling, and on the other hand, coiling

(a)

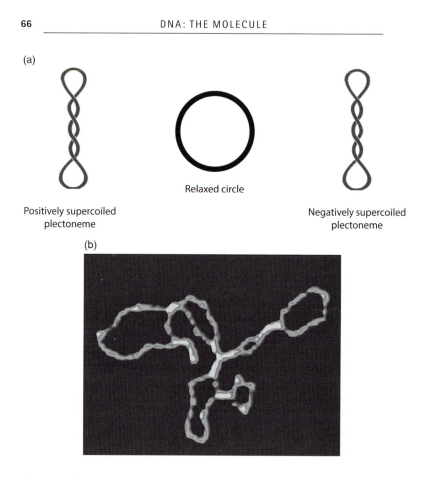

Positively supercoiled
plectoneme

Relaxed circle

Negatively supercoiled
plectoneme

(b)

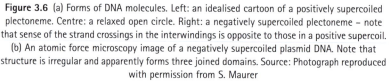

Figure 3.6 (a) Forms of DNA molecules. Left: an idealised cartoon of a positively supercoiled plectoneme. Centre: a relaxed open circle. Right: a negatively supercoiled plectoneme – note that sense of the strand crossings in the interwindings is opposite to those in a positive supercoil. (b) An atomic force microscopy image of a negatively supercoiled plasmid DNA. Note that structure is irregular and apparently forms three joined domains. Source: Photograph reproduced with permission from S. Maurer

can be changed by the movement under particular constraints of a protein, or protein complex, such as RNA polymerase along the DNA. An example of the first case is DNA gyrase, a bacterial topoisomerase that introduces negative supercoiling into DNA, thus facilitating both compaction and strand separation at biologically important DNA sequences. In this example, the energy level of DNA is raised. Other topoisomerases can reverse this effect, so relaxing DNA.

Supercoil Structure

The ability of DNA to be supercoiled is a fundamental aspect of its function and is intimately linked to the property that DNA is itself a coil – a right-handed double helix (Box 3.2).

In solution the coiling of a straight DNA molecule is described by the twist – the average angle between successive base-pairs. When the two ends of the molecule are fixed to maintain the length of the molecule and torque is applied, provided the molecule remains straight, the twist of the DNA will change. If the torque is in the same sense as the coiling of the double helix, i.e., right-handed, the twist will increase. Or, because the twist angle increases, the number of base-pairs for a 360° full turn of the double helix will decrease. Conversely, when the torque is left-handed the twist angle decreases uncoiling the double helix and ultimately facilitating the separation of the two DNA strands. However, because DNA is a flexible polymer applied torque need not be expressed simply as a change in twist. If the ends of the molecule are fixed but the DNA is free to bend, its helical axis can itself coil, resulting in the formation of a true 'supercoil'. In this state the DNA writhes. Importantly, between the fixed ends of a DNA molecule the writhe can be converted to a change in twist simply by re-establishing the

---

Box 3.2 DNA Handedness and Topology

Helices are ubiquitous, especially regular helices – a simple example in the living room may be a turned chair leg (often a right-handed double helix, while outside fire-escapes can be helical). All helices have a handedness – they are either right-handed or left-handed. This describes the sense of coiling. Right-handed helices – like DNA – are coiled in a clockwise sense and left-handed helices in an anti-clockwise sense. Helices can also vary in the number of strands. A protein α-helix has one, DNA and protein coiled-coils two, and the protein collagen three. Twisting described the number of units – e.g., base-pairs – in a single full turn of the helix whilst writhing describes the form of the 3D trajectory of the helix. This in turn is in essence helical although just as a topological circle can be much more irregular than the conventional concept of a circle so also with writhed helices.

Box 3.3 Interconversion of Twist and Writhe

A practical demonstration of a writhe to twist interconversion and vice versa is to take a short moderately rigid length of rubber tubing or hosepipe and hold it straight by holding both ends. Then, without letting go of either end, gently twist the tubing with one hand (the sense of the twist doesn't matter) for at least two turns. The tubing is now twisted. Gently relax the tension, still not letting go of the ends, and the tubing will writhe (you will feel it) first into a toroidal coil. Then with further relaxation the tubing will resemble the apical loop of a plectoneme connected to interwindings. The number of interwindings will depend on the number of turns of twist initially incorporated into the tubing.

---

straight DNA molecule by pulling (Box 3.3). Depending on the temperature, writhing is normally energetically favoured relative to twisting.

DNA Supercoiling

Supercoiling is an integral aspect of the biological function of DNA. In most bacteria and also in eukaryotic nuclei, DNA is negatively supercoiled – that is, it is in principle somewhat unwound relative to its energetically favoured state when free in solution. Much of this superhelicity is bound – or constrained – by protein scaffolds in the form of DNA coils in which much, if not all, of the superhelicity is in the form of writhe and packs the DNA into a smaller volume. Release of DNA from the scaffold allows the superhelicity in the molecule to re-equilibrate between twist and writhe, potentially decreasing the average twist.

The enzymatic manipulation of superhelicity can take two forms. On the one hand, an enzyme, termed a 'topoisomerase', can bind to a single site and directly change the coiling, and on the other hand, coiling can be changed locally by the movement under particular constraints of a protein, or protein complex, such as RNA polymerase along the DNA. An example of the first case is DNA gyrase, a bacterial topoisomerase that uses ATP to introduce negative supercoiling into DNA, thus facilitating both compaction and strand separation at biologically important DNA sequences. In this example the energy level of DNA is raised. Other topoisomerases can reverse this effect removing super-coils, either positive or negative, and so relaxing DNA.

However, while topoisomerases can establish and maintain an equilib-rium state of supercoiling, the processes of DNA replication and tran-scription generate transient changes in DNA supercoiling following the translocation of the protein complexes along the DNA. These transients arise as a direct consequence of the double-helical structure of DNA. When proteins such as RNA polymerase move along DNA, they do not track linearly along the molecule but instead by following one or other of the grooves rotate along a helical path. Many protein complexes are much more bulky than DNA, and their freedom to rotate around the DNA is constrained by molecular crowding. In some cases the polymerising enzymes may even be restrained in a fixed spatial position by physical attachment to extensive structures such as membranes. Under these circumstances, provided the rotation of the DNA molecule is also con-strained, torsional strain is generated such that the DNA is overwound downstream of the advancing enzyme and underwound upstream (Figure 3.7). This principle, first proposed by Liu and Wang (1987), plays a key role in the genetic organisation of chromosomes.

Another important function of topoisomerases is to buffer DNA structure against temperature variations. Living organisms exist over a broad temperature range of approximately –15°C to +120°C, and within this range individual organisms can tolerate quite wide temperature variations. At the high end of the overall temperature range the prob-ability of adventitious melting, and consequently of errors in, for example, transcription initiation, is substantially increased. To counter-act the possibility of such deleterious bubbles, extremophiles – for example, thermophilic bacteria and Archaea – often encode a reverse DNA gyrase that increases the twist of the DNA double helix. Indeed, recent calculations suggest such a strategy for stabilising the double helix would enable the DNA of an extremophile, *Thermus thermophilus*, to retain sufficient stability for biological function up to a temperature of 106°C or 15°C higher than the maximum value for free DNA observed in the absence of topological constraints. Even organisms that exist at more normal ambient conditions exert fine control over DNA structure to compensate for temperature changes. For example, the bacterium *Escherichia coli* maintains a constant superhelical stress over a

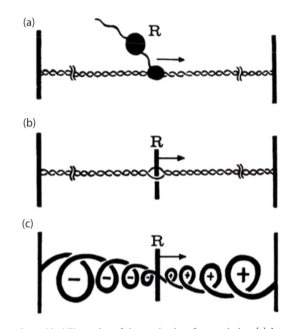

Figure 3.7 A graphical illustration of the mechanics of transcription. (a) A transcription ensemble, including polymerase, nascent RNA and proteins bound the RNA moving along a DNA segment, the ends of which are anchored. (b) The transcription ensemble acts as a divider separating the helical DNA into two parts. (c) If the polymerase moves from right to left without rotating around the DNA, the DNA in front of the polymerase becomes overwound or positively supercoiled, and the DNA behind the polymerase becomes underwound or negatively supercoiled. Source: Reproduced from *Proceedings of the National Academy of Sciences USA, 84*, 7024–7027, 1987, with the permission of the author, Liu & Wang, 1987

temperature range of 17–37°C. This effect, presumably mediated by topoisomerases, compensates for temperature-dependent alteration of double-helical pitch over this temperature range.

In both the bacterial nucleoid and the eukaryotic nucleus, DNA is usually packaged as a negative supercoil consistent with the preferential binding of negative supercoils by the core histones, most abundant nuclear DNA-binding proteins (the histones) and bacterial nucleoid (HU, H-NS and FIS) proteins. But why is DNA packaged in a higher energy state? Perhaps trivially, as in a ball of string, coiling is an efficient mode of packaging a long chain and so increases the compaction of long stretches of DNA. But the storage of negative supercoils also has the potential to facilitate the passage of DNA and RNA polymerases along a

DNA template. If DNA is already packaged as a negative supercoil, release of these negative supercoils could effectively neutralise some, but probably not all, of the positive superhelicity generated by an advancing enzyme and so counteract any inhibitory effect of positive torsion on the procession of the melted bubble within the polymerase complex. After the passage of the polymerase, the negative superhelicity behind the enzyme would facilitate the repackaging of DNA. Such a process would conserve negative superhelicity. The positive superhelicity could in eukaryotes be removed by relaxing topoisomerases, such as topoisomerase II, while in bacteria DNA gyrase could use ATP to ultimately increase the average level of negative superhelicity. This distinction highlights a fundamental difference between bacteria and eukaryotes. By possessing DNA gyrase, whose activity is sensitive to ATP levels, bacteria – as well as blue-green algae, and at least some mitochondria and chloroplasts – can fine-tune the negative superhelicity of the genomic DNA so that its energy level reflects energy availability from external sources. Lacking DNA gyrase in the nucleus, this mechanism is not available to eukaryotes. Nevertheless, they can, and do, by protein binding conserve the negative superhelicity generated by DNA translocases. An increase in unconstrained (not protein-bound) negative superhelicity on chromatin decompaction (Gilbert & Allan, 2014), could arise both from the release of constrained supercoils or from the selective relaxation of positive superhelicity generated by transcription.

If packaged DNA constituted an energy store for modulating gene expression, the nature of the packaging could depend on energy availability. In bacteria there is evidence that this is indeed the case. During the late stationary phase of growth when the cells are starved on energy, the nucleoid body collapses and the DNA is packaged by the highly abundant Dps protein. The mode of packaging by Dps is very different from that of the other abundant DNA binding proteins characteristic of exponential growth, and there is no evidence that it constrains DNA superhelicity, suggesting that high superhelical density is not a necessary concomitant of compaction. In eukaryotic nuclei, as in bacteria, loss of energy results in chromatin compaction. Here, however, the molecular nature of the mechanisms involved is not yet understood. Nonetheless, it seems a reasonable proposition that the coupling between energy availability and DNA organisation is tighter in bacteria than in eukaryotes.

Alternative DNA Structures

Although a right-handed double helix is the canonical image of a DNA, it has long been recognised that the molecule can adopt a number of other biologically important structures. These include variations on the double helix such as bubbles, Z-DNA, cruciforms and slipped loops, three-stranded triple helices and even distinct four-stranded structures, G-quadruplexes (Figure 3.8). The formation of all of these alternative structures is favoured not only by particular sequence organisations and base compositions but also by the energetic environment of these DNA sequences.

One of the simplest, and among the biologically most important, of these alternative structures is a DNA bubble in which strand separation occurs over a short sequence of base-pairs generating a region consisting of two separated strands bounded by more stable double-helical stretches. Bubble formation is strongly sequence dependent, occurring most frequently at the TpA base-step – the least stable of all 10 steps

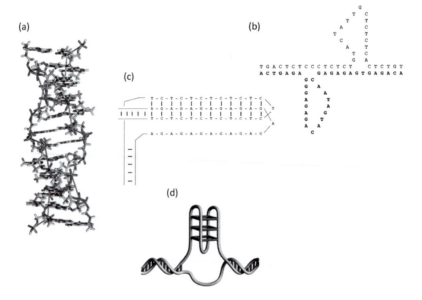

Figure 3.8 Folding of duplex DNA into alternative structures. Some structures, e.g., G-quadruplexes, may contain different modes of base-pairing from those in the standard double helix. (a) Slipped loop. (b) Cruciform. (c) Triple helix. (d) G-quadruplex.

Table 3.1 *Melting energies of the 10 base-steps (data from Protozanova, 2004)*

Base step	Melting energy (kJ/mole)
TA	−0.50
TG/CA	−3.26
AA/TT	−4.34
AT	−5.31
AG/CT	−5.39
CG	−6.01
GA/TC	−6.93
GG/CC	−8.23
AC/GT	−8.52
GC	−11.28

(Table 3.1). Consequently melting is favoured by agents – higher temperatures and negative superhelicity – that promote the unwinding of the double helix and the distribution of preferred melting sites closely correlates with genetic function. For example, DNA sequences in close proximity to the transcription startpoint are, on average, enriched in the TpA base-step, as also are specific recombination sites.

Negative superhelicity also facilitates the formation of other alternative DNA structures, notably the left-handed Z-DNA (Crawford et al., 1980) and also cruciforms and their slipped loop variant (Lilley, 1980; Mace, Pelham, & Travers, 1983; Minyat, Khomyakova, Petrova, Zdobnov, & Ivanov, 1995). Although there has been debate about the occurrence of such structures in vivo in some examples sequences with the potential to form these structures are located in the vicinity of promoter regions. G-quadruplexes constitute another class of structure (Sundquist & Klug, 1989). They form from a G-rich single strand with a particular sequence organisation and constitute the major structural motif at the single-stranded termini of eukaryotic telomeres. However, additionally such sequences are frequently found in internal positions, again often located close to promoter regions. Again, like Z-DNA and slipped loops, their formation is promoted by negative superhelicity.

The multiplicity of alternative DNA structures whose formation is dependent on the intrinsic torsional stress in the DNA begs the question of their function, if any. Their frequent association with promoter regions implies that they could facilitate, but not necessarily be essential for, transcription initiation. Although in textbooks this process is presented as relatively simple, in reality the progression from polymerase binding to the escape of an actively transcribing polymerase presents conflicting topological problems. After binding the enzyme first mediates the melting of approximately slightly more than one turn of double-stranded DNA, thereby constraining negative superhelicity. But in a closed system this melting of one double-helical turn must be balanced by the generation of an equal and opposite positive superhelicity, which will be most manifest in the immediate vicinity of the polymerase. In the absence of abundant topoisomerases relaxing this positive superhelicity, it could be absorbed by alternative DNA structures converting them back to a simple double helix. However, there is another potential barrier to initiation – the escape of the polymerase. If the DNA is initially relaxed as the polymerase begins to move away from the promoter, the advancing complex will generate negative superhelicity behind and positive superhelicity in front. Absorption of the negative superhelicity by a DNA sequence with the potential to form an alternative structure could then enable the release of the polymerase complex from the promoter. One possible function of these structures is thus to act as a torsional buffer. Similarly in eukaryotic chromosomes the positive superhelicity in front of polymerase could contribute to the unwrapping of DNA around nucleosomes.

Although the ability of DNA to assume a variety of alternative structures is a highly visible manifestation of conformational flexibility, the molecule can undergo other more subtle biologically important transitions. One such is the property of coiling or writhing under torsional stress. This is favoured by increasing flexibility and hence higher A/T contents and, importantly, can be induced by both negative and positive superhelicity. Negative superhelicity also induces localised strand separation, and in this case any choice between writhing and strand separation will likely depend on both the nature of the sequence and the prevailing environmental conditions. For example, because the untwisting of DNA increases as the temperature rises, in any topological constrained DNA molecule the partition between twist and writhe will change with temperature. Suppose a circular DNA is completely relaxed at low

---

Box 3.4 Topological Domains in DNA

A topological domain constrains topological transitions to a particular length of DNA and is operationally characterised by the limitation of rotation – sometimes to zero – about its two termini. This limitation in principle prevents any internal topological rotation being transmitted outside the confines of the domain. Topological domains can vary in length from as little as one turn of the DNA double helix (a microdomain) to many tens of kilobases. Topological domains were initially recognised in both bacteria and eukaryotes as functional units that contained a gene or groups of genes – in enteric bacteria the average length of a domain is ~10 kbp, while in eukaryotes domains can be much longer. Microdomains (Muskhelishvili & Travers, 1997) are generally short lengths of DNA defined by binding sites to a single protein or protein complex. Examples include RNA polymerase, recombination complexes and even nucleosomes where microdomains can be delimited by arginine contacts with the core histones spaced at intervals of one helical turn. A further property of domains is that they are often, but not always, transient, reflecting the dynamic nature of protein–DNA interactions.

---

temperature (say 4°), then as the temperature increases so more thermally labile sequences will untwist more and this (negative) untwisting will be compensated by a positive writhe in the remainder of the molecule.

The sequence-dependent bending anisotropy and elasticity of DNA have the potential to organise a DNA molecule in a preferred writhing configuration or configurations under superhelical stress. Such preferred configurations could contribute to the distinction of discrete domains within a single DNA molecule (Figure 3.6b). When stabilised by proteins such structures, which in an unconstrained DNA molecule are likely dynamic, could act as separate topological domains (Box 3.4) and enable differential modes of gene regulation within each domain.

The Importance of Torque in DNA Manipulation

So how do cells form bubbles in DNA (Box 3.1)? A simple way to visualise the process is to take a piece of double-stranded twine (preferably wound as a right-handed double helix like DNA, but it doesn't really matter) and then to close off a length of two to three helical turns by clamping with finger and thumb at each end. Then gently twist one end in the opposite sense to the helicity of the twine but still keeping it straight. The two strands will begin to come apart – they are untwisted – but the twine will retain its overall helicity. To form a bubble simply push one end of the clamped twine towards the other. If you're lucky a bubble will then form over part of enclosed region. The geometry of the bubble is remarkably similar to the structure of the DNA in a bubble that is primed for transcription for RNA polymerase (Figure 3.9).

So what's happening? Mechanically the clamped ends act as a wrench and allow the application of torque to the twine between them. The ends define a topological microdomain. Then the final push to form the bubble alters the trajectory of the unwound DNA. In this demonstration the torque is applied externally and so the closed domain containing the bubble can be distributed between a change in twist (the melting of the double helix) and a potential change in writhe (the change in trajectory) (Box 3.5). But how do the molecular machines responsible for transcription, DNA replication and recombination accomplish this transition? In the case of the multisubunit RNA polymerases responsible for most transcription in all cells the enzyme by virtue of its structure imposes a confined pathway for the DNA aligned almost at a right angle to the

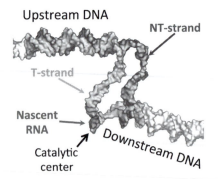

Figure 3.9 A DNA 'bubble' constrained by RNA polymerase. A similar structure can be obtained by manipulating a short stretch of double-stranded twine (see text). Note that the helical axis of the DNA is writhed. NT-strand and T-strand indicate non-transcribed and transcribed strands, respectively. Source: Reproduced with permission from Zuo and Steitz (2015), *Molecular Cell 58*, 534–540. Copyright 2015 Elsevier Inc. All rights reserved

Box 3.5 Twist, Writhe and Linking Number

Twist, writhe and linking number are the three geometrical properties that describe the topology of a closed DNA molecule such as a circle. In mathematics, the linking number is a numerical invariant that describes the linking of two closed curves in three-dimensional space. Because DNA has two strands, this definition translates to the number of times each DNA strand crosses the other. For a planar (relaxed) circle the linking number equates to the number of double-helical turns. Twist is a measure of the average rotation of one base-pair relative to its neighbours in three-dimensional space and is normally expressed as its reciprocal – the helical repeat – as bp/turn. Care must be taken to distinguish between two uses of twist – the intrinsic twist (and its reciprocal, the intrinsic helical repeat) and the surface twist (and its reciprocal, the surface helical repeat). The former, as its name implies, refers to the geometrical properties of a free DNA molecule, whereas the latter describes the geometry of a DNA molecule bound on a surface, usually a protein or protein complex, and is equivalent to the number of base-pairs between geometrically equivalent sites on a regular surface. Writhe for B-DNA is a descriptor of the three-dimensional path of the double helix. A planar circle has a net zero writhe but when DNA adopts a superhelical path writhe can be positive or negative depending on whether the superhelical coil is right-handed (clockwise) or left-handed (anti-clockwise). Both twist and writhe are quantitative measures of rotation. For most DNA allomorphs, twist is right-handed and therefore positive. For just a closed DNA molecule all topological transitions are defined by the equation:

$$\Delta Lk = \Delta Tw + \Delta Wr$$

where ΔLk is the change in linking number for whole system. Because Lk is invariant, this means that any change in Tw must be accompanied by a corresponding opposite change in Wr.

In both bacteria and eukaryotes the topology of DNA is tightly controlled. In many bacteria the DNA is negatively supercoiled – that is, it contains fewer right-handed turns than the corresponding relaxed DNA molecule, i.e., Lk is lower. Such changes are mediated by a suite of enzymes, termed topoisomerases, which change the coiling of a DNA molecule by introducing or removing superhelical turns.

trajectory of the entering DNA. This pathway takes the form of an enclosed channel and contains both the bubble itself and the site at which RNA synthesis occurs.

Although this structural transformation seems to be fundamentally a simple process, in the cell, it is often much more complicated. The limiting factor is the origin of the torque to apply to the wrench. In principle tight contacts between the enabling enzyme, for example, RNA polymerase, and the DNA at a promoter site may provide sufficient binding energy. In certain cases, for example the promoters of some bacteriophages (bacterial viruses), this situation is likely the case. However, a uniform application of this principle would preclude many possibilities for direct and indirect regulation – a consequence that would be anathema to the operation of a complex system. Many bacteriophage promoters just need to fire rapidly for the virus to take over the bacterial cell, and so there is no obvious reason why they need to be regulated. They represent the exception that proves the rule.

But if the binding energy is insufficient – for transcription initiation the jargon is that the promoter organisation is suboptimal – what is the source of the torque that enables bubble formation? This additional torque requires energy. It can take the form of either the ATP-dependent DNA gyrase or the progress of, say, RNA polymerase utilising the amount of available unconstrained DNA negative superhelicity, usually in the form of writhe. Subsequently polymerase may trap this writhe or cooperate with certain transcription factors that transduce this superhelicity to the polymerase wrench. Ultimately this process again represents a writhe to twist conversion. Alternatively DNA melting may be actively promoted by transcription factors that use ATP to impose a torque on and stabilise a bubble. Such proteins are found not only in bacterial and eukaryotic transcription, but also in DNA replication initiation complexes.

In all these processes the fundamental topological principle is that in a closed domain the melting of the double helix where the change in twist is negative is accompanied by a compensating change in writhe, i.e., the writhe becomes more positive (Box 3.6). In technical terms ΔWr is positive and ΔTw negative. (Note that these considerations only apply to closed constrained systems and not necessarily to test-tube experiments using small fragments of DNA in an open unconstrained system). The crucial requirement for DNA melting in a closed system is that ΔWr is positive.

This could be the result of a negative writhe becoming less negative, or of a negative writhe changing sense to become a positive writhe, or again of a positive (or zero) writhe becoming more positive. This principle is well illustrated again by the bacterial RNA polymerase (Amouyal & Buc, 1987). Simplifying what is in practice a rather complicated process, current evidence indicates that when negative supercoiled promoter DNA first binds to RNA polymerase the enzyme traps writhe by wrapping the DNA in a left-handed superhelical loop (Figure 3.10). As the polymerase progressively forms a melted bubble, the superhelix compacts, becoming less writhed and may ultimately become a right-handed superhelix.

But this simple description begs the question of the origin of the necessary torque, or more precisely, how is the force applied externally to the enzyme? The geometry of the left-handed superhelical loop wrapped around the polymerase of a plectonemically supercoiled DNA circle. Such apical loops, which occur naturally in supercoiled DNA, would constitute appropriate binding sites for the free enzyme so that the enzyme would be located at an apex of the supercoil immediately adjacent to the interwindings of the plectoneme. With this organisation a change in the wrapped superhelical writhe could be effected by a right-handed rotation of the loop relative to the interwindings so that the writhe of polymerase-wrapped DNA became more positive to accommodate the untwisting.

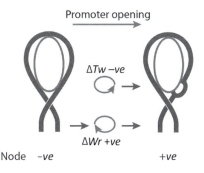

Figure 3.10 Model for how negative superhelicity drives promoter opening. Wrapping of apical loop DNA around RNA polymerase allows formation of untwisted bubble accompanied by a change in the sense of the first crossing (node) of the plectonemic interwindings.

Utilising negative superhelicity in this manner is an option available to certain bacteria, especially common gut bacteria, but in yeast, a simple eukaryote, the RNA polymerase responsible for the transcription of mRNA, this mechanism is apparently not available. Yet, like the bacterial enzyme, the yeast enzyme also wraps DNA but with one important difference – the sense of the wrapping is not left-handed but right-handed. In this situation it is conceivable that bubble formation by the eukaryotic enzyme increases the right-handed writhe of the wrapped DNA.

Transcription initiation is one example of how a biological system can utilise torque. Another is the unwrapping of nucleosomes both to decompact chromatin and to facilitate the passage of nucleosomes by a transcribing RNA polymerase. In a nucleosome the DNA is packaged as ~1.6 turns, mostly writhe, of a left-handed supercoil. Unwrapping to, say, a linear DNA molecule is then accompanied by loss of negative writhe, i.e., the change in writhe (ΔWr) is again positive. But how to facilitate this process? Both the nucleoid of bacteria and the nuclei of eukaryotes contain abundant DNA-binding proteins that untwist DNA so that the untwisted DNA is held – constrained – within the binding site of the protein. In bacteria these are the HU proteins and in eukaryotes the HMGB proteins. If any of these proteins bind in a closed topological domain, this change in twist (ΔTw is negative) will be compensated by a positive change in writhe. This implies that when, for example, an HMGB protein binds to wrapped nucleosomal DNA, provided neither it nor the DNA is free to rotate, just like a key in a lock, the protein will apply positive torque to the nucleosomal DNA and initiate unwrapping. In biological systems the DNA binding domain is often part of a more complex protein or protein complex that binds to nucleosomes. One example would be the so-called pioneer transcription factors involved in triggering cell fate and sex determination. Another is the complex, FACT, for facilitating the passage of a transcribing RNA polymerase through an array of nucleosomes. This requires the partial unwrapping – loosening – of the wrapped DNA coupled with the removal of two of the eight histone proteins around which the DNA is initially coiled. A third example is provided by the protein complexes that shuffle nucleosomes along the DNA. Some, but not all, of these again loosen the wrapping and contain multiple copies of the HMGB domain. The property of inducing positive writhe – or reducing negative writhe – is not restricted to abundant chromosomal protein. It is also a characteristic of a eukaryotic major transcription factor, TATA binding protein aka TBP. This property likely contributes to the formation of the transcription bubble at TBP-dependent promoters.

DNA and Genetic Organisation

The physicochemical properties conferred by DNA sequence not only determine bending and melting preferences but also strongly correlate with the genetic organisation of both eukaryotic and bacterial chromosomes. In general the coding sequences of genes have a G/C-rich bias. In part this is because the codons for the most abundant amino acids also have a G/C-rich bias. The corollary is that non-coding DNA sequences, including introns as well as 5′ and 3′ flanking DNA sequences, are generally more A/T-rich. Indeed the most A/T-rich and most thermodynamically unstable DNA sequences in the *Saccharomyces cerevisiae* genome are located in 3′ flanking regions. This distribution of base composition on a genomic scale implies that, on average, coding sequences are stiffer or less bendable while non-coding sequences are both more flexible and more susceptible to strand separation. However, in apparent contradiction to these variations in flexibility, in eukaryotic chromosomes coding sequences have a higher nucleosome occupancy than non-coding sequences. But, again, this pattern of occupancy is possibly related to the higher intrinsic entropy of A/T-rich sequences.

The occurrence of the more A/T-rich sequences in the flanking regions of genes has functional significance. At the 5′ end of a transcription unit there is an obvious correlation with the requirement for RNA polymerase to melt DNA prior to transcription initiation. But at the 3′ ends of transcription units polymerase dissociates and releases the constrained unwound DNA so that it reforms a double helix. One possibility is that such regions serve as topological sinks, absorbing by writhing any positive superhelicity generated in advance of the transcribing enzyme. This would block the transmission of any such superhelicity to a neighbouring gene and the potential disruption of its chromatin structure. Instead the writhed DNA would serve as an appropriate substrate for relaxation by topoisomerases and so establish a domain boundary.

The relationship of the physicochemical properties of DNA to chromosome organisation and function is not only apparent at the level of individual genes and transcription units but is also a feature of whole bacterial chromosomes. These chromosomes comprise, in general, a single circular DNA molecule that can vary in length from ~0.5 Mb to 6–10 Mb. Remarkably in these chromosomes, at least in most γ-Proteobacteria, gene order is highly conserved such that those genes that are expressed highly during exponential growth are clustered near the origin of DNA replication, while those that are more active during episodes of environmental stress resulting in the cessation of growth are more frequent in the vicinity of the replication termini. However, not only is there a gradient of gene organisation from origin to terminus but also this gradient correlates, on average, with a gradient of base composition so that in each replichore the most stable, G/C-rich, DNA is close to the origin while the least stable is at the terminus. This average pattern of course includes wide variations at the level of individual genes. Yet another feature that exhibits a graded response from origin to terminus is the distribution of binding sites for DNA gyrase, a topoisomerase that inserts negative superhelical turns into DNA. Again these are concentrated primarily in proximity to the origin of replication and thus create the potential for the DNA in this region to be more highly negatively supercoiled than that close to the terminus. This overall pattern of organisation can couple chromosome structure to energy availability. When bacteria are shifted to a fresh rich growth medium, ATP levels rise activating DNA gyrase and thus increasing the negative superhelical density of the chromosome. This would be localised to the origin-proximal region and would in turn activate the genes producing the necessary components for growth – the transcription and translation machinery – as well as providing an appropriate environment for DNA replication. Once DNA replication is initiated the passage of the replisomes along the two replichores would by itself generate a gradient of superhelicity by the Liu/Wang principle with the more negatively supercoiled DNA again being located closer to the origin and the more relaxed DNA close to the terminus. Again, by analogy to transcription, the DNA close to the terminus, in concert with topoisomerases, could act as a topological barrier between the two replichores. The bacterial chromosome thus functions as a topological machine where the overall distribution of DNA sequences reflects the coupling between the processing of the replisomes and gene expression.

Why DNA? The Emergent Properties Conferred by DNA Superhelicity

DNA superhelicity has long been regarded as a rather esoteric and peripheral property of the DNA molecule. Yet it may be argued that it is one of the major factors contributing to the rise of DNA as the principal genetic material. In an evolutionary perspective it can be regarded as an excellent example of an emergent property that operates as a substrate for natural selection. Emergent because the generation and maintenance of superhelicity require both that the DNA molecule has a double-helical character and also that the rotation of the molecule about its long axis be severely constrained either by circularisation or by the physical attachment of the end of the molecule to a rigid support. Without these two conditions any superhelicity generated by chance would be rapidly dissipated.

In the context of modern organisms, how does DNA superhelicity contribute to the operation of the genetic programme of a cell? It is an analogue property than can vary over a range of very approximately 1 positive coil per 100 base-pairs (about 10 double-helical turns) to 1 negative coil per 100 base-pairs and changes both the helical repeat and writhe. On average cellular DNA contains 1 negative coil per 4–6 double-helical turns of which most is in the form of packaging by DNA binding proteins. The stored energy in these negative coils can potentially be made available to facilitate transcription and DNA replication and thus play a direct role in regulating gene expression.

Not only might the relatively static packaged coils act as an energy store but also the ability of supercoils to propagate along a molecule potentially enables the coordination of expression of neighbouring genes. For example, if two neighbouring genes are transcribed in opposite directions from a single intergenic region – they are divergent – then the negative superhelicity generated upstream of a transcribing enzyme on one gene has the potential to kickstart the transcription of the neighbouring gene by actively facilitate the melting of DNA at the transcription startpoint. This capacity for intergenic communication mediated by the DNA superhelicity generated by RNA polymerase can thus contribute substantially to the evolution of the local gene organisation on chromosomes. This is a very similar phenomenon to the global shaping of the genetic organisation of whole chromosomes by the passage of DNA replisomes.

Although the physicochemical properties of DNA can determine its dynamic roles in the context of enzymatic manipulation, how are they integrated with the primary function of DNA as an information store? Like DNA, a fundamental property of RNA is to encode protein sequences, but unlike a DNA genome an RNA genome must combine both information storage and its functional expression during the translation of the nucleotide sequence into protein. These two requirements are not necessarily wholly compatible. For example, when there are strong selective pressures to maintain the integrity of the genomic nucleotide sequence, the option of regulating translation by, for example, modulating the half-life of an RNA molecule is effectively excluded. Perhaps more tellingly, in known present-day biological systems, RNA molecules that function both as messengers and as genomes (for example, those of poliovirus and Qβ bacteriophage) are, relative to most genomic DNA molecules, very short, comprising only a few thousand nucleotides. Crucially these genomes exist primarily in the single-stranded form and thus lack the essential double-stranded characteristic supporting supercoiling. RNA genomes also differ from DNA genomes in one other important respect. When an RNA genome is replicated, the molecule is copied from one end generating a double-stranded form, which then serves as a template for the synthesis of multiple copies of single-stranded RNA genomes. However, in most cases the replication of DNA starts at an internal site and then proceeds outwards in both directions simultaneously. This mechanism retains the overall topological structure of the molecule.

An advantage of the separation of responsibilities between the two types of polynucleotide is that the juxtaposition and catenation of individual genes into much longer molecules enhances the potential regulatory repertoire of gene expression. In particular the coordination of gene expression can be facilitated at the local level by the structural interplay arising from the transcriptional activity of adjacent genes, depending on whether they are organised in tandem or transcription is convergent or divergent. At a higher level of structural organisation the continuity afforded by a single DNA double helix in a chromosome permits the organisation of genes, and hence the available DNA information, into distinct structural and functional domains comprising many protein-coding elements. Apart from these considerations, the ability of DNA to accrete more and more packets of information – as genes or as regulatory elements – into longer and longer chromosomal molecules is arguably an important factor contributing to increases in organismal complexity. Any increase in the length of chromosomal DNA molecules should be coupled to – and possibly limited by – mechanisms for the generation and maintenance of genomic integrity. In this context a relevant biological example is provided by ciliates – a group of single-celled organisms including the causal agents of malaria and sleeping sickness – where regulation of DNA-directed gene expression is separated from maintenance of the germ line. These organisms contain two types of nuclei. One, the micronucleus, containing the complete diploid genome, serves as the germ line and does not express genes, while the second type, the polyploid macronucleus, contains a highly edited and fragmented version of the genome and serves as the vehicle controlling gene expression. Only the micronucleus undergoes mitotic chromosomal segregation.

The emergence of DNA superhelicity as a major mode of packaging DNA and of coordinating gene expression provides powerful arguments for evolutionary scenarios. Because of its double-helical structure the movement of DNA translocases along the molecule would likely generate superhelicity – perhaps initially small amounts, but with greater resistance to rotation, larger amounts would accrue. Translocation by itself is topologically neutral – that is, the number of negative supercoils produced balances the number of positive supercoils. But in the present day within a nucleus there is an excess of negative supercoils principally associated with abundant DNA-binding proteins such as histones that bind negative, but not positive, supercoils. Selection for such proteins would confer two advantages – more stable packaging and the storage of DNA at a higher (negative) energy level that would enable the passage of translocases. In this scenario the positive supercoils are less stable and preferentially lost while the energy stored in the negative supercoils can be regarded as a by-product of polymerase progression. Put another way, the superhelical character of DNA was initially dependent entirely on translocase activity. That could still be the case in the modern-day nucleus as there is no described alternative mechanism for converting energy in the form of ATP into DNA superhelicity.

DNA Packaging

A DNA molecule is essentially a linear device for information storage. As such the amount of information contained in a single DNA molecule can be huge (Box 3.6).

While the genomes of the simplest bacteria with a single chromosome contain very approximately 5×10^6 bp, the genome of the Japanese 'canopy plant', *Paris japonica,* distributed among several chromosomes contains 1.5×10^{11} bp. This latter value is about 50 times the DNA content of a human cell. Given that within a duplex individual base-pairs are separated on average by 0.34 nm, this diversity translates into genome lengths of only 0.0017 m of DNA in the nucleoids of simple bacteria and 100 m in *P. japonica,* with our own genome measuring about 2 m. The immense length and corresponding potential information content of individual DNA molecules poses a conundrum. How can such an extremely long and flexible chain be packaged into the very small volume of a bacterial nucleoid or a eukaryotic nucleus and, at the same time, be organised in such a manner so that access to the enzymatic machinery responsible for transcription and DNA replication is maintained? To put this question into a starker perspective, in actively growing bacteria the DNA packing density in the nucleoid has been estimated to be ~80–100 mg/mL, while in bacterial spores the density can be at least 10 times higher. Comparable densities occur in eukaryotic nuclei. These high-packing densities imply that, on average, the DNA duplexes are very close together. In the most compact

---

Box 3.6 Information Content of DNA

A gigabyte of information corresponds to 4×10^9 bits and, making the simplest (and undoubtedly inaccurate) assumption that the information content, 2 bits, of every base-pair in a DNA molecule is equivalent and independent, a 2×10^9 bp DNA molecule would contain a gigabyte of information.

assemblies of DNA molecules there may be as little as 2.4 nm distance between the centres of individual duplexes, meaning that the space between the duplexes is ~0.4 nm (Figure 3.11) (Leforestier & Livolant, 2009, 2010; de Frutos, Leforestier, & Livolant, 2014). This contrasts with the average dimensions of multi-subunit RNA polymerase molecules of approximately 10–12 nm. This conceptual problem of maintaining access while storing a lot of information in a very small volume has obvious analogies to the retrieval of information stored on a computer hard disc.

In biology two intrinsic properties of DNA provide a solution to the conundrum of efficient access to stored information. First, under appropriate conditions DNA can condense into extremely compact structures, and second, DNA supercoils possess a structural variety that enables the genome to be marked with distinguishable tags. Together these properties confer structural organisation on a DNA genome. More fundamentally, the optimum structural requirements for the packaging

Figure 3.11 The Japanese 'canopy plant', *Paris japonica*, contains the largest recorded genome size in angiosperms. *A black-and-white version of this figure will appear in some formats. For the colour version, refer to the plate section.* Source: Photograph in the public domain, reproduced from Wikimedia

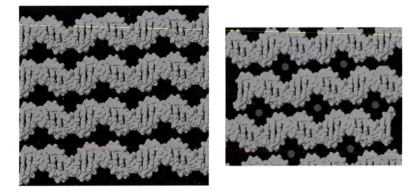

Figure 3.12 Modes of close packing of DNA duplexes within a bacterial virus particle. Left: the duplexes are aligned such that major and minor grooves are opposed. Right: the alignment when the major groove of one duplex is opposite the minor groove of its neighbour. The average distance between the helical axes of neighbouring duplexes is little more than the duplex width of 2 nm. Source: Reproduced with permission from Leforestier and Livolant (2009), *Proceedings of the National Academy of Sciences, 106*, 9157–9162. Copyright 2009 National Academy of Sciences

of a DNA molecule that is being actively replicated and transcribed conflict with those where, as in certain virus particles or metaphase chromosomes, the DNA duplexes are stacked as irregular planar arrays (Figure 3.12), an organisation that effectively minimises the volume occupied by the DNA. It is this difference that provides a key insight into the origins of multicellular complexity.

The essential difference between a DNA molecule that is actively involved in transcription or replication and one that is maximally packaged is that whereas in the former case the DNA motion is essentially rotational, in the latter such motions would be likely incompatible with tight packaging.

Chromosome Structure and Organisation

Today DNA genomes exist in a variety of physical forms. Chromosomes can be circular or linear, and the DNA molecule can be both single- or double-stranded, although the former is very much the exception. To some extent these differences correlate with the complexity of the

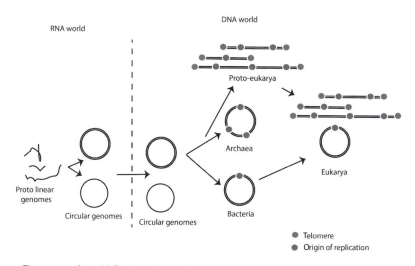

Figure 3.13 A model for the evolution of genomes. In Eukarya linear nuclear chromosomes coexist with circular mitochondrial and also, in plants, chloroplast genomes.

organism. Bacteria and Archaea usually have circular chromosomes, most often a singleton, whereas more molecularly complex organisms have multiple linear chromosomes (Figure 3.13) But why? Adaptation of genomes in environments in flux requires the generation of sufficient genetic variation for natural selection to act. The evolution of genetic mechanisms, and so the DNA organisation, that facilitate the production of variation are crucial. One such major step was likely the emergence of linear chromosomes with protected ends – telomeres. This innovation arguably has significant consequences for the evolution of biological complexity.

First, the step change from circular to linear chromosomes is ultimately correlated with the total amount and hence potential information in an organism's genome. Both circular and linear chromosomes vary substantially in length. Circular chromosomes, found in bacteria as well as mitochondria and chloroplasts, range in size from a paltry 11,000 bp to ~1,000,000 bp. In contrast, linear chromosomes can attain average lengths of up to ~40,000,000,000 bp in organisms as diverse as *Paris japonica* and the lungfish *Protopterus aethiopicus*. Yet in reality the size range of eukaryotic chromosomes is vast. For example, the length of the smallest in the budding yeast, *Saccharomyces cerevisiae,* is a mere

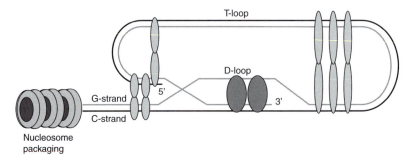

Figure 3.14 Organisation of telomeric DNA. The 'G'- and 'C'-strands are folded in a looped structure stabilised by proteins. The single-stranded extended terminus of the G-strand base-pairs with a short region of the C-strand to form the D-loop.

270,000 bp. Although both types of chromosome contain a single DNA molecule, in circular chromosome, unlike linear chromosomes, the two DNA strands are continuous, but in contrast linear chromosomes are capped by structures termed 'telomeres' in which the two strands at the end of a DNA duplex are staggered with respect to each other so that *telomeric* DNA contains double-stranded tandem repeats of TTAGGG followed by terminal 3' G-rich single-stranded overhangs. These form a complex loop sealed by associated proteins (Figure 3.14).

Both the capping of linear chromosomes by telomeres and the circularity of small bacterial chromosomes point to a common selective mechanism – the preservation of genetic and hence informational – integrity of a DNA molecule. Compared to a linear DNA molecule with no capping a circular DNA molecule is protected from being disabled by the complete separation of its component strands. Consider the effect of a transient rise in temperature above the upper melting point of the DNA in a chromosome. In a linear chromosome the strands would completely separate and be unable to rapidly reanneal and, perhaps worse, might interact with similar sequences from other chromosomes. For a DNA circle, however, the continuity of the strands retains the topological relationship between them, so ensuring that reannealing to reform the DNA helix will occur immediately on the restoration of the ambient temperature. Similarly telomere capping by protecting the ends of the chromosomal DNA molecule would preclude unraveling of the double helix from the end. In both cases the absence of exposed ends would prevent, or at least minimise, any nibbling from the ends mediated by enzymatic hydrolysis.

Although linear chromosomes are clearly correlated with longer DNA molecules, there are likely both costs and benefits. The rate of DNA replication is relatively constant so for a single DNA replication protein complex the amount of time needed to replicate a DNA molecule would in principle depend on the length of the molecule. The initiation of DNA replication in bacterial and archaeal small circular chromosomes generally occurs at a single location, although in some archaeal chromosomes multiple origins occur and then proceeds bidirectionally around the molecule until both replicating complexes reach the termination site situated directly opposite to the origin. Because the maximum rate of DNA synthesis is effectively fixed, a single origin restricts the size of a circular chromosome. The more DNA the chromosome has, the longer it will take to be replicated. For rapidly growing organisms like Bacteria this is potentially a problem. The time needed to replicate the chromosome may exceed the doubling time of the organism. For a bacterium such as *Escherichia coli*, the doubling time is 20′ but the time required to replicate a whole chromosome is 40′. The imbalance is circumvented by the single origin firing twice during a generation. However, because the chromosome is a closed circle, multiple firing of a single origin is likely to increase the probability of DNA tangling within the chromosome. In contrast to bacterial and Archaeal chromosomes, the linear chromosome of eukaryotes contain multiple origins of replication, which can be simultaneously or sequentially activated. The potential tangling problem is eliminated. Nevertheless in simple eukaryotes, such as budding and fission yeasts – the multiple individual chromosomes are of comparable length to circular bacterial chromosomes. It is only in higher eukaryotes that the expansion of the genomic DNA content is accompanied by substantial increases in both the length of the chromosomes and the gene content.

A further potential advantage of linearity is that it allows the alignment between two copies of the same chromosome in a diploid nucleus during cell division (mitosis). Similarly, alignment during the process of meiosis (Box 3.7) necessary for the formation of germ cell provides a mechanism for the reassortment of genetic material. Because the two chromosomes carry different allelic variants, this meiotic reassortment increases the variety of possible allelic variations in the haploid germ lines and therefore the adaptive possibilities in any resultant diploid progeny (Box 3.7).

Box 3.7 Meiosis

Meiosis is the process in which the diploid chromosome content of the fertilised zygote (containing two copies of each chromosome, one from each parent) is rearranged to yield germ cells containing only a single copy of each chromosome.

Box 3.8 Dinoflagellates

Dinoflagellates are a divergent eukaryotic evolutionary group comprising approximately 2,000 extant species. They are predominantly marine, and some species are responsible for toxic 'red tides'. The group contains the endosymbiotic Zooxanthellae, which inhabit various coelenterates (including coral-building forms) and protists, as well as certain molluscs.

Increasing the range of adaptive possibilities is not the only consequence of linearity. Arguably linearity may also promote biological complexity. In meiosis chromosomes pairing occurs at multiple sites along their length. With closed circular chromosomes – typical of Bacteria and Archaea – multiple pairing, and the resultant intertwining of two circles, would create a major topological problem. With telomeres chromosomal identity becomes distinct and is associated with both localisation and chromosomal multiplicity within the nucleus. Coupled to the possibility of increasing the length of DNA molecules in linear chromosomes is the potential for increasing the number of genes and, hence, information contained within a genome.

The step change in chromosome structure between bacteria and eukaryotes is an emergent property contributing to the stability of longer chromosomes and hence more DNA information. Simplistically more DNA information potentially allows the further evolution of complexity, but there is little, if any, correlation between, say, multicellularity and genomic DNA content (Raff & Kaufmann, 1983). For example, some single-celled dinoflagellates (Box 3.8) have a DNA content closely comparable to that of *Paris japonica*.

4

The Evolution of Biological Complexity

In my view all salvation for philosophy may be expected to come from Darwin's theory.
—Ludwig Boltzmann, 1905

. . . complexity means that a plurality of elements, here actions, can be linked only selectively.
—Niklas Luhmann, 1984

The Nature of Complexity

Many discussions of the biological world use the terms *complex* or *complexity*. But what is the real sense of these words? Conceptually the usage of *complexity* is slippery and is often used synonymously with *diversity*. In any system consisting of many components, its diversity, in the sense used in this book, is a measure of the number of different types of entity present and does not, on its own, necessarily imply any interactions between them. The components of a system could be entirely independent. However, complexity is a different beast and is admirably exemplified by biological systems. In a complex system the different components interact with, and influence the behaviour of, other components of the system. In this sense the complexity of a system is a measure of a functioning whole. To be complex, a system is necessarily diverse because the greater the diversity the greater the number of possible interactions between the individual parts of the system. In a more general sense, complexity is a measure of interactions in a system of heterogeneous components.

Luhmann's definition of complexity in the context of a 'system', although arising from a discussion of social systems, is equally valid when considering physical and, by extension, biological systems. His 'actions' describe a network of interconnecting pathways of communication in, his example, a society. In principle these are functionally equivalent to a network of molecular interactions in a cell and to a network of direct, but less intimate, interactions in an ecological system. In this way of thinking, it is possible to dissect a 'system' into a number of subsystems, which together constitute the whole but which themselves may properly be regarded as systems. Here the definition of a system distinguishes between what is internal and what is external. Internally the logic of the network defines its operation, whilst signals from the external world may modulate the functioning of the internal interactions.

Such general discussions beg the question, what is the nature of a biological system? To some extent a biological system may be defined, for the purposes of discussion, simply in terms of the number of interacting components required to facilitate a particular function. For example, a particular chemical reaction may be catalysed by a single biological macromolecule – an arbitrary complexity of 1. If, however, the rate of this reaction can be changed – increased or decreased – by the direct molecular interaction of the catalysing biomolecule with a second biomolecule, then in the context of biological function the complexity of the system is increased to an arbitrary 2. However, inside a cell such simple systems are likely to be a rare exception rather than the rule. Many proteins have the ability to interact directly with many other proteins – not necessarily all simultaneously – which in turn also can interact with many other different proteins. The result is a network of interactions that hone necessary chemical reactions so that the rates of production of metabolites are fully integrated with the functioning of the cell.

Such networks are characteristic of cells. But many organisms are multicellular, and the functioning of the organism as a whole requires communication between the component cells. This is accomplished in diverse ways – by the diffusion or directed transport of small-molecule messengers from one cell to another and also by direct molecular interactions between cell surfaces. Once again the proper functioning of the system – in this case a multicellular organism – is dependent on a network of interactions and hence represents a further increase in complexity over a single cell.

This much is perhaps obvious, but organisms do not live in isolation. They exist in communities – ecosystems – and almost by definition ecosystems are maintained by a network of interactions, many not necessarily involving intimate molecular contacts, between the component organisms. The question is then whether ecosystems are simply a different manifestation of biological complexity as a whole, or is the formal nature of the complexity distinct? If the formal nature is different, it should be possible to define a dividing line between ecosystems on the one hand and individual organismic systems on the other. In practice, any such line is likely to be arbitrary and, as in any attempt to define distinct categories in biology, to be riddled with exceptions.

Simplistically the biological world may be thought of as a hierarchy of systems of which together describe an overall system of increasing complexity, with each higher level of complexity incorporating a number of less complex systems. At each level an essential requirement is communication between the included pre-existing systems of lower complexity. Thus the increased molecular complexity consequent on the emergence of DNA required communication with pre-existing RNA-based systems to maintain information processing. Such increased communication might be manifest as an increased number of components of protein complexes – again an increase in complexity. Within individual cells the development of distinct compartments or organelles with distinct functions again requires increased communication between different systems within the cell and the acquisition of the necessary information in the genome to encode the mechanism of communication. Again the development of multicellularity implies communication between different types of cell in an organism.

A possible, if perhaps misleading, mathematical analogy to the hierarchy of biological complexity in which each level contains a variety of networked systems would be the Mandelbrot set in which intricate patterns recur at every level of magnification. However, while the Mandelbrot set is infinite and essentially homogeneous, in contrast a hierarchy of biological complexity is finite, heterogeneous – perhaps more akin to a delta.

A legitimate question is to what extent these differences in complexity are reflected in the amount of DNA – the genome size – of an organism. In multicellular organisms one common measure of apparent complexity is the number – the diversity – of different cell types. Here the consensus is that there is no direct correlation between genome size and the number of different cell types. Even within individual groups of organisms – one might consider arthropods or angiosperms – there can be an extremely wide range of genome size for similar degrees of cellular diversity. Nevertheless, simple organisms that lack internal membranes separating functionally distinct regions – for example, extant bacteria and Archaea – generally possess small genomes ranging from ~0.5 Mb to 5 (Archaea) – 20 (bacteria) Mb. Similarly the genome size of fungi and sponges is intermediate between on the one hand that of bacteria and Archaea and on the other the average genome size (~2,000 Mb) of most other eukaryotic divisions. The variation in the amounts of genomic DNA obscures differences in gene organisation and number, but at the most basic levels of organisation there remains a positive correlation between genome size and cellular complexity. The lack of such a correlation with cell type complexity implies that any required additional information is not directly manifest in the primary DNA sequence but itself is a property of additional information storage complexity.

For any networked system the robustness of a process depends on an ability to respond or adapt to local environmental fluctuations. The responses of many biological systems, at whatever level, behave as though they are adapted to and buffered over a defined range of fluctuation. For example, the gut bacterium *Escherichia coli* grows efficiently over a range of about 20°C between about 20°C and 40°C. Outside this range survival is possible but is more severely compromised the greater the departure from optimal growth temperatures. Robustness can be increased by overlapping functional redundancies in the network, which allow the system to exploit a wider range of fluctuations. In essence, this increases the complexity of the system because the number of components and interactions within it also increases. In turn, the inclusion of overlapping redundancies dependent on more components means that the information available to the system has also increased. In this example, more complexity requires more information, and for cells and organisms this information is encoded in the DNA. An analogous argument can be made for eco-systems. In this case functional overlapping redundancies in the net-work make take the form of, for example, a variety of predators that may feed on similar ranges of prey but have different preferences. Or again, within the system there may be a variety of organisms promot-ing the breakdown and reutilisation of organic material with each organism having preferred substrates. These preferences are selective and again encoded in the DNA. Again increasing complexity correl-ates with increasing information content. However, in both cells and ecosystems functional redundancies are not independent. Rather, they interact with each other and consequently the response of the system as a whole results in the operation of an optimal pathway or pathways for a given environmental condition. This is simply a restatement of the notion that complexity arises from the interaction of discrete heterogeneous elements.

Another way of thinking about the issue of biological complexity is to consider the nature of the processes involved. The biological world overall is essentially a system in constant flux; that is, for its mainten-ance it depends on a continual flow of chemical reactions. It is dynamic, and this dynamism requires energy. This consumption of energy is balanced by the production of small molecules – for example, carbon dioxide and water – in an 'open' thermodynamic regime. In this context the operation of a biological system – however defined – then depends on the availability and utilisation efficiency of the rele-vant energy source, provided that other necessary conditions for life as we know it are met.

Within a cell the principal primary energy currency is adenosine $5'$ triphosphate (aka ATP). This molecule is required for the production of polynucleotides (RNA and DNA) and also of proteins. It also, directly or indirectly, drives the dynamic directed (non-random) motions of molecules within the cell. For molecular systems of simple complexity, even if the chemical reactions catalysed by the system are energetically neutral or quasi-neutral, ATP is required for the pro-duction of the component macromolecules. At the cellular level with a higher order of complexity, the primary energy source, at least in animal cells, is derived from small molecules, such as sugars, pro-duced by the breakdown of larger biological macromolecules. These are oxidised to produce the cellular supply of ATP. In plant cells, which contain chlorophyll as well as single-celled green algae and blue-green algae – the latter being a type of bacterium – the primary energy source is sunlight. Photons are harvested by light-absorbing pigments, and the energy is ultimately used to oxidise water with the concomitant formation of oxygen in the process of photosynthesis. This transformation is then coupled to the fixation of carbon dioxide to form sugars.

More complex multicellular organisms differ in their primary energy source. Animals rely exclusively on biomolecules, while for (most) plants the primary source is light. Yet only for those plant cells that contain chlorophyll is light the primary source. Others – root cells are the most obvious example – lack chlorophyll and instead rely on the sugars produced by photosynthesis. The difference in primary energy source between animals and plants means that, with the possible exception of more exotic energy sources for carbon fixation, sunlight is the ultimate energy source for all biological systems on the planet. Within an ecosystem energy availability is dependent on light capture, but the efficiency with which that energy can be utilised is likely dependent on the recycling of energy sources by the system. Essentially this implicates energy as a critical factor the efficiency of the regeneration of biomolecules to forms that can be utilised by other organisms in the ecosystem. In simple physical terms, energy availability is a function of the intensity and duration of the light source while the output, representing very crudely the overall efficiency of energy conversion, is the biomass of the system, such as a rainforest, being considered.

Selection for efficiency of energy utilisation implies that the totality of the multitudinous chemical reactions within a biological system be finely tuned by mutation to provide a selective advantage. This does not mean that in chemical terms the reaction conditions are necessarily optimised, but that the conditions are good enough or slightly superior when compared with those in competing systems. Fine-tuning in turn implies that the activity of individual reactions be tightly regulated. Within cells this is accomplished by increasing the direct interactions between molecules. This manifestation of increasing complexity effectively reduces the intrinsic entropy of the system relative to systems with no such interactions and requires increased information content. The succeeding sections discuss how this process might operate at the molecular level.

The Genetic Code

The evolution of the genetic code may be considered to be a simple paradigm for the evolution of biological complexity. The current view is that the code evolved from a simple two-letter non-overlapping form in which two successive nucleotides specify an amino acid to a three-letter form, where the specification can require three successive nucleotides. Associated with this process has been the preservation of distinct code variants where certain triplets encode different amino acids from those found in the great majority of organisms. A further development is that the specification of at least one amino acid requires more information than a single triplet.

The evolution of the code is a phenomenon for which there is plethora of hypotheses and a dearth of data. However any hypothesis must ultimately be consistent with the thermodynamics of the coding process. Central to this process and, more importantly, a quantifiable parameter, is the interaction of an RNA codon in messenger RNA with its complementary anticodon in transfer RNA. During biological evolution these interactions almost certainly predated the adoption of DNA as the genetic material. The correlations between codon–anticodon stability and present-day coding variability argue in favour of the evolution of divergent genetic codes accompanied by a transition from a two-letter triplet code to a three-letter triplet code.

When the base step stability of the first two bases in a codon–anticodon pair is compared to coding capacity, several correlations are observed (Table 3.1). In general, the most stable initial base steps correspond the coding boxes containing a single amino acid in the standard code, while the less stable ones are associated with boxes containing two or even three outcomes. This correlation is dependent on melting stability and not on stacking stability alone. The difference in melting energies between the base steps GT/AC and GG/CC and also between CT/AG and AT is sufficiently small that the constituents of these pairs, given the discussed uncertainties, are essentially equivalent. Second, decreasing melting stability can be accompanied by an increase in the complexity of the polypeptide chain, for example, the length of the carbon chain associated with a particular amino acid, and a decrease in the chemical stability, of the encoded amino acids. Finally, chain termination is directed by the least stable steps and in the present-day translation machinery does not require base-pairing. Instead in the absence of a tRNA a protein – a 'release' factor – specifies and facilitates protein chain termination.

This analysis of the standard code neglects the phenomenon of coding variability where the same codon can encode different amino acids in different organisms In general the greatest differences in coding capacity occur between the standard code, and non-plant mitochondria, while significant, although less extensive, differences are found between the standard code and that utilised in ciliates and certain Mycoplasma species. Strikingly, when the variability of coding capacity for particular codons is compared with the stability of the first base step, the most extensive variation is associated entirely with the five least stable base steps (Table 3.1). Notably, some of the least stable triplets exhibit the greatest variability. For example, in different contexts TGA can encode termination, cysteine, tryptophan or selenocysteine, while TAA can encode termination, tyrosine or glutamine. In contrast the coding capacities of the five most stable base steps in the first two positions of a triplet encoding alanine, glycine, valine, proline, glutamic acid, aspartic acid and arginine are essentially invariant.

These correlations just described are consistent with a simple model for the evolution of the genetic code (Figure 4.1). As suggested by both Jukes (1963) and Crick (1968), the initial code would have been a two-letter triplet of the form XYNXYN... in which the codon–anticodon interactions would have been mediated by the XY base step. A coding box XYN would on this model specify a single amino acid. Subsequently a three-letter triplet utilising the so-called wobble rules would have developed generating a code of the form XYZXYZ... so that a coding box could specify two or more amino acids (see Box 4.1 for an explanation of base-pairing and 'wobble'). Interestingly, the CT/AG base step, which has an intermediate melting energy, specifies one coding box with a single amino acid and a second with two (Table 3.1). In this scenario the specification of an amino acid by a particular codon would depend largely on the availability of the cognate amino acid, accounting for the inverse correlation of base step stability with the complexity and chemical instability of the encoded amino acids. Further, the conservation of coding capacity for the most stable base steps would be consistent with the view that codes that differ significantly from the standard code could have arisen by divergent evolution at a stage when only a few amino

Second letter

First letter		U		C		A		G		Third letter
U		UUU UUC	Phenylalanine	UCU UCC	Serine	UAU UAC	Tyrosine	UGU UGC	Cysteine	U C
		UUA UUG	Leucine	UCA UCG	Serine	UAA UAG	Stop	UGA UGG	Stop Tryptophan	A G
C		CUU CUC	Leucine	**CC**U **CCC**	Proline	CAU CCC	Histidine	**CG**U **CG**C	Arginine	U C
		CUA CUG	Leucine	**CC**A **CC**G	Proline	CAA CAG	Glutamine	**CG**A **CG**G	Arginine	A G
A		AUU AUC	Isoleucine	ACU ACC	Threonine	AAU AAC	Asparagine	AGU AGC	Serine	U C
		AUA AUG	Isoleucine Methionine	ACA ACG	Threonine	AAA AAG	Lysine	AGA AGG	Arginine	A G
G		GUU GUC	Valine	**GC**U **GCC**	Alanine	GAU GAC	Aspartic acid	**GG**U **GG**C	Glycine	U C
		GUA GUG	Valine	**GC**A **GC**G	Alanine	GAA GAG	Glutamic acid	**GG**A **GG**G	Glycine	A G

Figure 4.1 The triplet genetic code. The triplets are arranged as normally presented. The simultaneous occurrence of G and/or C in the first two positions of a codon is indicated in bold.

Box 4.1 Base-Pairing and 'Wobble'

The term 'wobble' base-pair was coined by Francis Crick (1966) and describes a base-pair formed between two nucleotides in RNA molecules that does not follow the classic Watson–Crick base-pairing rules and so is not part of a uniform double helix. Initially it was applied to explain the pairing associated with the third base-pair of a codon–anticodon interaction but has subsequently been observed in other RNA secondary structures. The four main wobble base-pairs are guanine–uracil (G–U), hypoxanthine–uracil (I–U), hypoxanthine–adenine (I–A) and hypoxanthine–cytosine (I–C), where I represents inosine, the nucleoside containing the base hypoxanthine. By utilising a wobble base-pair for recognition of the third base in a codon two different codons, for example, UGU and UGC encoding tyrosine, can pair with a single anticodon within a tRNA molecule.

acids were specified. An alternative view that non-standard codes arose entirely by evolution from a fixed standard code is not supported by the observed pattern of variability. Nevertheless, the adoption of a novel coding capacity from the standard code would be consistent in some examples with the notion that the lower the energy of the codon–anticodon interaction the greater the possibility that this coding interaction could be usurped by an interloper. In this context two relevant codons are those for tryptophan (UGG) and methionine (AUG). In the modern canonical code these amino acids are the sole examples of coding by a single codon. In both the first two nucleotides specify relatively less stable interactions in codon-anticodon pairing.

The amino acid selenocysteine is structurally similar to cysteine, but contains selenium instead of sulphur. Although it is one of the least abundant amino acids in proteins, it is an important constituent of the active sites of certain enzymes. It is widely, but patchily, distributed in the biological world, being found in bacteria, Archaea and eukaryotes. Selenocysteine has its own tRNA (Sec-tRNA^{Sec}) but lacks a dedicated codon in the standard genetic code. Instead, cells employ a specialised mechanism for its integration at a dedicated recoded UGA stop codon. The encoding of selenocysteine requires not only UGA but also a structural element (usually a hairpin) in the mRNA. The decoding involves the formation of a canonical codon-anticodon complex of UGA with Sec-tRNA^{Sec} but also the interaction of a specific elongation factor, which binds both to the Sec-tRNA^{Sec} and also to the mRNA hairpin.

The evolution of the genetic code and its variants provide an example of how a gradual increase in stability of codon–anticodon interactions, by whatever mechanism, could extend the amount of accessible information that is available to a biological system. Initially the RNA copying would produce molecules containing all four bases, A, C, G and U, probably in different proportions. But at the start the ability to form a stable and productive decoding interaction equivalent to a codon–anticodon interaction would be restricted to double-helical structures with the most thermally stable doublet of base-pairs. This would favour G/C-containing doublets leaving the more A/T-rich sequences lacking a coding function, except perhaps chain termination. But with an increasing ability to stabilise codon–anticodon interactions – equivalent to an actual or effective lower temperature – these A/T-rich sequences would become available for productive coding. The consequence of this process would be that the coding capacity of the primordial mRNA would become more efficient and predictable. More efficient because a greater proportion of the molecule would be available for coding and more predictable because the more stable codon–anticodon interactions would be less susceptible to error. Put another way, the increase in stability at lower effective temperatures reduces the intrinsic Gibbs/Boltzmann entropy, and this in turn is accompanied by an increase in the potential coding capacity of the information carrying molecule. Where the lowering of the effective temperature is insufficient to provide an acceptable predictability, one or more additional intermolecular contacts provide the required stabilisation energy. In other words, more information – that encoding the additional contact(s) – is required for an increase in complexity.

The evolution of the genetic code not only illustrates how coding capacity can be elaborated by the successive stabilisation of less stable codon–anticodon interactions, it also highlights the constraints of the system. The issue is, why does the standard genetic code specify only 20 amino acids? The code is consists of triplets that are contiguous and non-overlapping. Because RNA essentially contains only four different bases, the possible number of triplets is 64. However, if indeed in the initial stages two letters and a space – i.e., XYN where N can be any base – were sufficient to specify an amino acid, then the available number of doublets or dinucleotides is only 16. An early assignment of a particular doublet to an amino acid would thus lock up 4 triplets and correspondingly reduce the potential number of amino acids that could be incorporated into the code. And indeed, with only one interesting exception, the 9 most stable doublets, accounting for 36 triplets in total, each encode a single amino acid. Only when the interaction of the third codon base with its partner in the anticodon became sufficiently stable would it be possible for an XYN block to become differentiated so that it could encode at least two, or possibly more, amino acids. Because the most stable codon–anticodon interactions had already been claimed, these additional codons would, on average, form less stable codon–anticodon pairs. A further limitation is the base-pairing between the third base of the codon and the corresponding anticodon. Here there is more spatial flexibility in the base-pairing rules than in a standard double helix and consequently a loss of precision. The base-pairs so formed do not necessarily correspond to the standard Watson–Crick base-pairs. The 'wobble' as originally proposed by Francis Crick supposes that the third base in a codon can pair with the corresponding anticodon base such that G in the 5′ anticodon position can pair with either U or C as the 3′ codon base. But C in the 5′ anticodon position can only pair with G as normal. U in the 5′ codon position is even less specific, being able to pair with A, G and also U. A further complication is that bases in the 5′ anticodon position in the tRNA are often chemically modified by enzymes, and these modifications can themselves affect the selectivity of the interaction with the codon. This loss of precision imposes further restrictions on the coding capacity so that, in general, each XYN box codes for no more than two amino acids, one with U or C in the third position and the other with A or G. These considerations suggest that the initial establishment of the functioning machinery for protein synthesis limited the extent to which it can be elaborated and hence the coding capacity of the DNA. Indeed, although there are 64 triplets (or 61 if the termination codons are excluded), the number of different tRNA molecules in a cell is sometimes no more than 45.

The interesting exception is the case of aspartic and glutamic acids. In the present-day codes, these amino acids are specified respectively by GAU/GAC and by GAA/GAG. Notably this pair of amino acids are also the most chemically similar in any of the coding boxes, and both are significant products of the early experiments to imitate a prebiotic environment. Although a code of the form $(XYN)_n$ has, in principle, the capacity to specify only one entity, the entity specified does not have to be unique. Put another way, in terms of the initial specification of amino acids, the code could be sloppy with all the GAN codons directing the incorporation of either aspartic acid or glutamic acid. Only subsequently, possibly with the evolution of the 'wobble' rules, would discrimination between these two amino acids have evolved.

This view of the evolution of the code raises the question as to why the code in its initial form should have been of the two-letter triplet form rather than of a two-letter doublet form. This issue was addressed by Crick (1968), who pointed out that if an initial RNA adaptor had a stem loop structure with the anti-codon on the loop, then the width of the stem would be ~20 Å. In A-RNA the P-P distance ~6 Å. The binding of two adaptors to successive codons requires contiguity of both codons and amino acid linkage sites. Consider two types of doublet code: an uninterrupted form XYXYXY... and an interrupted form XYNXYNXYN. For a doublet interaction of XYXY type, distance between successive codons is ~6 + (8–9) = ~14–15 Å; for a doublet interaction of XYNXYN type distance between successive codons is ~6 + 2(8–9) = ~22–24 Å. Only in XYN case is there sufficient distance in mRNA to accommodate simultaneous binding of two adaptors – unless there is a sharp change of direction in polynucleotide chain, which could abrogate contiguity of amino acid sites. This problem would be even more acute if there were originally three adaptors lined up, as in present-day ribosomes. Even if the primordial mRNA were aligned on a positively charged surface a code of the form XYXYXY... would still be excluded.

The evolution of an increasingly complex coding capacity is paralleled to a degree by an increase in the molecular complexity of the encoded amino acids. The amino acids specified by the most stable doublets in an XYN triplets are generally simpler than those encoded by the less stable doublets. Indeed, the amino acids with six or more carbon atoms are found encoded exclusively by XY(N) doublets with the lowest melting energies. Extreme examples are glycine and alanine encoded by the more stable GG(N) and GC(N) doublets, respectively, where N can be A, G, C or U, while tryptophan, the most structurally complex amino acid, is encoded by UG(G) and tyrosine, another aromatic amino acid, by UA(U, C). UA and UG are the least stable of all doublets in codon–anticodon interactions. Clearly the number of carbon atoms in an amino acid is not the only measure of its complexity. Other features include the adoption of chemical innovations – a sulphydryl group in cysteine and carboxamide side chains in asparagine and glutamine.

A necessary condition for the incorporation of an amino acid into the early genetic code is that it pre-exist in the environment and that therefore its production can be effected by a non-biological chemical mechanism. Classic experiments have shown that amino acids can be generated by an electric spark discharge, and among the major products are glycine and alanine, consistent with their early adoption into the genetic code. However, an apparent exception to the temporal order of codon acquisition is arginine. This amino acid was not found among the products of a spark discharge but is encoded by all four codons in the CGN box and so was, on the preceding arguments, possibly an early component of the code. The question then becomes whether the early prebiotic synthesis of arginine is chemically plausible. This amino acid has an unbranched 5-carbon chain linked to a guanidinium group. Interestingly, a major product of the discharge experiments, only exceeded in abundance by glycine and alanine, was the unbranched 5-carbon amino acid norleucine. This amino acid is not a current constituent of proteins. The generation of a guanidinium group could be facilitated by effective concentrations of ammonia and cyanide derivatives such as cyanamide and dicyanamide. Both hydrogen cyanide and dicyanamide have been invoked as significant primordial reactants for the production of biological precursor molecules. From this perspective, two possible scenarios can be invoked: either the primordial concentration of arginine was sufficient for it initially to be encoded as such or, possibly less likely, CGN first coded for a different 5-carbon amino acid such as norleucine and subsequently this amino acid was replaced by arginine. Whatever the mechanism was that facilitated the allocation of CGN to arginine, once this had occurred, there could have been considerable selective pressure for its maintenance. An alternative explanation is of the amino acids specified by a single XYN box, arginine is the only basic one and is today found both in single-stranded RNA binding motifs and in DNA packaging proteins including histones. Arginine containing protein molecules, as well as peptides such as protamine, could enable DNA condensation as well as constrain the flexibility of coding RNAs, possibly in a particular configuration, and so facilitate their translation and/or their enzymatic activities. Such structural stabilisation would then constitute a platform for further selection.

Sequence Complexity: Did the Evolution of the Genetic Code Accompany More Precise Molecular Recognition?

An evolutionary expansion of the genetic code has profound implications for the precision of macromolecular recognition and for the information content of a biological system. Corollaries to this expansion could be a change in the coding content of RNAs from a G/C-rich to a more uniform base composition and a decreasing intrinsic stability of the codon–anticodon interaction. One consequence of progressively more amino acids becoming available for coding purposes is that the products of an initial codon-driven translation would have been dominated by peptides containing only four amino acids – for the sake of argument, glycine, alanine, proline and arginine, all of whose codons have G or C in the first two positions. Such a composition would not only have placed serious limitations on the secondary structure of the peptides but also longer sequences containing just these amino acids would have a tendency to be repetitive. Many similar repetitive amino acid sequences – aka SLiMs or short linear motifs, MFs, etc. (Van Roey, 2014) – are highly abundant in contemporary genomes, but it is not known what fraction of such sequences are involved in phase separation. Not only do the proteins containing them perform essential functions but also structurally the repetitive protein sequences are often distinct from the classical concept of a compact well-folded protein. Instead by themselves they are highly mobile and are intrinsically disordered, having no recognisable defined structure (Wright & Dyson, 1999; Watson & Stott, 2019). DNA-binding proteins often contain sequences of this character. For example, the eukaryotic chromatin-binding histone protein, H1, contains a long C-terminal tail dominated by proline, alanine and arginine and, to a lesser extent, serine. And in the bacterium *Pseudomonas*, a very similar DNA-binding protein, AlgR3, is composed largely of tetra- or penta-amino acid repeats containing only alanine, proline and lysine (Medvedkin et al., 1995). The association of these peptides and DNA is largely driven by charge neutralisation of the basic residue (in this case lysine) with the phosphate backbones of the DNA strands. Both the DNA and protein sequences are repetitive, so the sequences cannot by themselves define a unique complex between the interacting components. Instead when bound the polypeptide chain can remain highly mobile and disordered (Turner et al., 2018) but still effects DNA compaction.

With the accession of more available amino acids to the genetic code the number of possibilities for protein sequence could, on a combinatorial basis, be immensely expanded as also would be capacity to form more ordered three-dimensional structures. Such sequences are more information rich and encode more ordered three-dimensional structures often stabilised by internal interactions mediated by hydrophobic residues, the majority of which are specified by codon–anticodon interactions of lower thermal stability (Travers, 2006).

An intrinsic ability of some repetitive peptides of limited amino acid composition to facilitate the formation of condensed, yet dynamic, structures – coacervates (Box 4.2) – is a key component of

---

Box 4.2 Coacervates

Coacervates are ubiquitous components of cellular biological systems that separate different parts of a cell into distinct liquid phases acting as functional compartments (reviewed by Hyman, Weber, & Jülicher, 2014). A simple analogy would be an emulsion. The concept has a venerable history. Coacervates were first recognised by E. B. Wilson in 1899 as globules in the cytoplasm of starfish eggs. Different phases can vary substantially in extent and may, for example in the interior of a chromatin fibre, occupy a very small volume. In cells they are often highly dynamic both within and between phases where rapid exchange of biomolecules often occurs. The boundaries between distinct liquid phases may or may not be delineated by a visible structural barrier. Within the eukaryotic nucleus the nucleolus is a distinct compartment but is not separated from the remainder of the nucleus by an obvious barrier. Similarly, the DNA compartment in eukaryotes – the nucleus – is separated from the remainder of the cell by a membrane, whereas in prokaryotes there is no membrane surrounding the DNA compartment – the nucleoid – although it is usually spatially delimited.

---

compartmentalisation into separated but communicating liquid compartments in a small volume. This allows not only functional separation but also the selective concentration of different reactants in different compartments. If you like, in a biological system, not unlike in present-day industrial practice, different chemical factories have distinct locations. Coacervation is also a crucial facet of DNA function and results in distinct compartments within the larger nuclear compartment. In evolutionary terms is it coincidence that the simple repetitive sequences facilitating DNA condensation and coacervation differ only in one essential basic amino acid (lysine in place of arginine)? But even now the repetitive sequences in some of the more stable condensed chromatin structures between DNA and chromatin histone proteins contain both arginine and lysine as the dominant basic amino acids. Because the prevailing ionic environment is also a determinant of the stability of peptide–DNA complexes, it is conceivable that an initial selection of alanine, proline and arginine for a peptide engaging with nucleic acids might not have been dependent only on availability but also on an early – and different – ionic environment.

The liquid–liquid phase separations driving compartmentalisation are selective in that the direct chemical interactions – even if only transient – differ between compartments. Again the example of nucleic acid–peptide interaction is simple and not readily generalisable to a similar binding of a basic repetitive peptide to other macromolecular components. Of course it is likely that the nature of the short repetitive sequences could be changed by evolutionary pressures, but it remains striking that the current chemical organisation of the cell could in principle have an extremely ancient heritage. Nevertheless, by definition, a simpler and repetitive peptide would have a lower information content than longer, more varied sequences. Similarly the complexes between such sequences and their ligands would, by virtue of the redundancy in possible interactions, be imprecise and again have a lower information content than, say, highly specific sequence recognition. Thus inevitably the expansion of the genetic code would result in an increase in information content and complexity.

Macromolecular Complexity

An increase in the efficiency of biological catalysis dependent on the association of two or more macromolecules is ubiquitous in biological systems. A very simple example is the complex of an RNA molecule and a protein molecule that is involved in producing a functional tRNA molecule from the initial transcribed RNA copy of a tRNA gene. For most, if not all, tRNA genes, when the gene is copied by RNA polymerase the tRNA sequence is embedded in the longer RNA transcript and can only become functional after this transcript has been processed. In bacteria the first step in this processing – the removal of the RNA sequence immediate preceding the tRNA sequence – is performed by the enzyme ribonuclease P (RNase P). This enzyme is unusual in that it contains one protein chain and one RNA chain, and both are necessary for activity in the cell. However, in the test tube the RNA molecule can catalyse RNA cleavage in the absence of the protein, but the protein increases the rate of cleavage although it itself has no catalytic activity. How does it do this? Cleavage requires the recognition of the RNA substrate by the catalytic RNA molecule and a sufficient stability of the resulting binary complex to allow sufficient time for cleavage to occur. The accessory protein increases the binding affinity for the substrate RNA and in so doing would increase the probability of cleavage by the catalytic RNA. Put another way, the increased complexity provided by the presence of two, rather than one, components in the enzyme shifts the distribution of the intrinsic entropy in this simple system.

In the context of biological evolution RNase P is ancient. The enzyme is found in all kingdoms of life – Bacteria, Archaea and Eukaryota – as a complex of RNA and protein. In all these cases it is the RNA molecule that possesses catalytic activity. But comparisons of the enzymes from the three kingdoms illustrate a common trend in the evolution of molecular complexity. Whereas the bacterial enzyme contains a single protein molecule associated with a single RNA molecule in Archaea, RNase P, although again containing a single RNA molecule, is associated with 4–5 different protein molecules, each of which is individually dispensable for the processing of a tRNA precursor. In the eukaryotic nucleus RNase P is even more complex, 9–10 protein molecules being associated with a single catalytic RNA molecule. Of these proteins five are related in sequence to Archeal RNase P proteins. However, there is a caveat to this simple pattern of an increase in molecular complexity with evolution, if only because there are other possible molecular solutions to the problem of RNA precursor processing. For example, the need for this form of processing would be eliminated by simply starting the transcript at the first nucleotide of the mature tRNA molecule. Or cleavage could be accomplished without the catalytic RNA molecule. Indeed, this actually appears to be the case for RNase P in chloroplasts and in animal mitochondria where the RNase P complex consists only of protein.

Despite such complications – possibly inevitable in a biological world – increases in molecular complexity with evolutionary progression are characteristic of many enzymatic activities. Among the most notable of these is the ribosome – the platform for the translation of mRNA into protein. Like RNase P, the ribosome is assumed to be evolutionarily ancient and like RNase P is a ribozyme – its essential catalytic activity, that of peptide bond formation, being mediated by one of its component RNA molecules. Ribosomes consist of two large macromolecular assemblages termed the small and large subunits.

In bacteria the small, 30S, subunit contains 1 RNA molecule of about 1,540 nucleotides associated with 21 proteins, and the large, 50S, subunit contains 2 RNA molecules, 120 and 2,900 nucleotides in length, associated with 31 proteins. In contrast, in eukaryotes the RNA in the small subunit is longer (~1,900 nucleotides) and is associated with more (33) proteins. The eukaryotic large ribosomal subunit displays a similar pattern. It contains 3, rather than 2, RNA molecules (120, 160 and 4,700 nucleotides in length) associated with 46 proteins. Archeal ribosomes, like bacterial ribosomes contain three RNA molecules but are associated with a variable number of proteins in the range of 50–70, presumably reflecting the diversity of Archeal lineages. However, although the RNA complexity is similar to that of Bacteria, the functionality of the ribosome more closely resembles that of eukaryotes.

Another example of an increase in molecular complexity associated with evolutionary progression is found in RNA polymerase, the principal RNA synthesising enzyme in the cell and consequently a major conduit for regulating information flow. There are strong structural and functional homologies between the RNA polymerases of Bacteria, Archaea and Eukaryota, but again they differ in the number of polypeptide chains in the functional complex. The bacterial enzyme contains 6 chains, whereas the archaeal enzyme contains 12 chains. In eukaryotes there is again an increase in the molecular complexity of the transcribing enzymes. Instead of the single RNA polymerase found in bacteria and Archaea, eukaryotes possess three distinct nuclear RNA polymerases. Each of these enzymes has a dedicated function: RNA polymerase I transcribes ribosomal RNA in the nucleolus, a sub-compartment of the nucleus, RNA polymerase II transcribes messenger RNA, while RNA polymerase III transcribes principally short RNA species. Thus not only is the increase in complexity manifest in distinct functions but also in the occupation of distinct subcellular locations. The three eukaryotic RNA polymerase are generally similar in structure and related to the Archeal enzyme each containing 12 subunits, some of which are common to all the enzymes. However, in addition to the separation of functions, one polymerase, RNA polymerase II, has a radical structural innovation. In contrast to other polymerases, one subunit has acquired a long extension (the CTD) to its C-terminal end. This extension contain up to 52 repeats of a hepta-amino acid sequence Tyr-Ser-Pro-Thr-Ser-Pro-Ser and serves as a landing platform for other proteins, particularly those involved in the processing and maturation of mRNA. These include enzymes required for the capping of the 5′ end mRNA transcript, proteins involved in the directed splicing of the transcript and yet others required for mRNA termination. Further functions of the CTD include the regulation of RNA chain initiation and the chromatin structure. All these different activities serve to coordinate transcription with cotranscriptional events dependent on the passage of polymerase II along a gene. Transcription is coupled directly to the post-transcriptional processing of the RNA product and to changes in chromatin structure in front of and behind the transcribing enzyme. And this coupling is facilitated by the direct recruitment by the CTD of the necessary protein factors to RNA polymerase itself. All this constitutes a remarkable illustration of the operation of complex molecular system – direct communication between a substantial number of proteins resulting in the direct coordination of the complex processes of transcription, mRNA processing and chromatin remodelling. Yet the association of different proteins the CTD can itself be regulated, adding yet another layer of complexity to the information processing system. Several of the amino acids in the heptad repeat can be modified by the addition of a phosphate group. Different modifications enable the recruitment of different proteins, and hence different functionalities, to the CTD. Crucially the state of modification of the CTD depends on the position of the polymerase on the CTD. Initially phosphorylation accompanies the initiation of RNA synthesis, both releasing the enzyme and enabling the polymerase to recruit the necessary additional proteins. As the gene processes through a relatively static polymerase, the pattern of modification changes so that spectrum of recruited proteins changes in concert with the functional requirements of the processes to be coordinated. The activities of enzymes that effect this phosphorylation are themselves regulated and ultimately depend on signals emanating from the metabolic state of the cell.

The coordination of different processes by the CTD of RNA polymerase II exemplifies the capacity of biological complexity at the molecular level to integrate processes and signals – a paradigm of organisation. But it is important to appreciate that this single example talks to and is integrated with other molecular systems responding to some of the same and some different signals. This overall organisational complexity is perhaps the hallmark of eukaryotes. For example, eukaryotic polymerase II requires more auxiliary general transcription factors than the corresponding archaeal enzyme. Although separate from the polymerase, these factors are integral components of transcription machinery necessary for RNA chain initiation. Nevertheless, although a relatively greater complexity distinguishes eukaryotes from other organisms, the converse – a decrease in complexity – needs to be considered. The transcription of simple genomes – for example, in some viruses and in mitochondria – is mediated by much simpler enzymes. These may contain only a single polypeptide chain but possess sufficient functionality to initiate transcription from specific DNA sites.

RNase P, the ribosome and RNA polymerase are but three instances of many exemplifying an increase in the number of entities associated with a particular function as the complexity of organisms increases. But RNase P and RNA polymerase also illustrate the converse. In much less complex systems – for example, viruses or those that are evolutionarily primordial – the molecular machinery can be correspondingly much simpler. It is conceivable, for example, that originally RNase P comprised only the RNA moiety. Correspondingly one of the simplest RNA polymerases is encoded by the small bacterial viruses, bacteriophages, T3 and T7. These viruses have a small DNA genome of 40,000 base-pairs, which contains sufficient information to allow the rapid multiplication of the virus in a bacterium. This process results in the ultimate lysis of the host and the release of mature virus particles. The viral RNA polymerase copies only viral genes into RNA, although its own mRNA is copied from the viral DNA by the RNA polymerase of the host. Unlike the host polymerase, the viral enzyme is effectively unregulated having the sole function of transcribing viral mRNA species required for the generation of more virus particles. In this example functional simplicity equates to molecular simplicity.

Increases in the complexity of molecular assemblies raise two issues. How do they evolve, and why, or in what situations, is complexity favoured? RNase P, because of its relative simplicity, is a good starting point. Whilst it is clearly impossible to know the precise sequence of evolutionary events, plausible scenarios for increases in molecular complexity exist. For RNase P we assume that the initial state was an RNA molecule with the ability to cleave other RNA molecules at certain positions along their chain. It was a simple RNA enzyme or ribozyme. This function could depend on the precise three-dimensional structure of the molecules to be cleaved. However initially the cleavage process might have been inefficient and lacked specificity, perhaps because the RNA chain of the ribozyme was so dynamic that the reactive groups in the chain only rarely approached each other sufficiently closely to catalyse the reaction or perhaps because the interaction with the substrate RNA molecule was weak and the resulting complex was not sufficiently stable to support an effective reaction rate. In both these cases, although they are clearly not mutually exclusive, the ribozyme functions inefficiently. How to increase efficiency? One possibility – the present-day situation for bacterial RNase P – would be to associate with another entity – which could be an RNA or a protein molecule – that would stabilise the structure of the catalytic RNA molecule, so reducing its wriggle rate, and/or the complex between the ribozyme and its substrate. The selected solution in bacteria is association with a small protein that enhances the reaction rate. This could have arisen by a favourable mutation in one of a large number of similar primordial polypeptides. Provided that the enhanced reaction rate constituted a favourable change in the system once the association of the ribozyme with a second component had occurred, then it would be the binary complex of protein and RNA, and not the original catalytic RNA alone, on which natural selection could act for the further evolution of the system.

One consequence of the scenario in which a binary complex sup-plants a single entity is that the amount of effective information encoded in the genome is immediately increased because two genes rather than one are required. A second consequence is that the stabilisa-tion of the ternary complex between ribozyme, auxiliary protein and substrate implies a decrease in intrinsic entropy of the system. Effectively the protein acts as a chaperone (Box 4.3 and Figure 4.2) to constrain the RNA in a particular configuration. Both the ribozyme and the protein will likely be more flexible – have a higher wriggle rate – when apart than when together. The same applies to the association of the substrate RNA molecule to the binary enzymatic complex. Most importantly, the utilisation of a binary complex implies an increase, if only small, in the organisation of the system.

There are necessarily reciprocal interactions between the ribozyme and the associated protein. The further evolution of such a system could involve the accretion of more proteins into the complex as seen in

Box 4.3 Molecular Chaperones

Chaperones are a ubiquitous feature of molecular biological systems. They enable the rapid adoption of preferred configuration of a single macromolecule or macromolecular complex, or in some cases they can act to stabilise a particular conformation of, say, a macromolecular motor. A universal example is provided by enclosed 'pots', which facilitate protein folding in a protected environment. When a cell is challenged by an abrupt rise in temperature – aka 'heat shock' – many proteins unfold, a process accompanied by enhanced production of protein-folding chaperones. Ultimately this process restricts the number of possible configurations adopted by a protein or protein-DNA complex by increasing the number of components in the system.

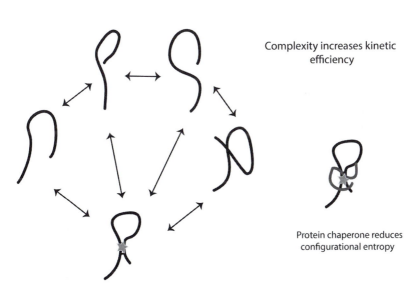

Complexity increases kinetic efficiency

Protein chaperone reduces configurational entropy

Ribozyme with no chaperone

Figure 4.2 Role of a protein chaperone stabilising an active configuration of a ribozyme. The active centre is marked by a star.

Archaea and Eukaryota with a corresponding increase in the number of inter-entity, normally inter-protein, interactions. Some of these would be internal to the complex, but others could communicate with other proteins external to the immediate RNase P system.

The preceding argument implies that the evolution of the more complex binary system could provide a platform for the evolution of greater complexity. The initial scenario was couched in terms of an increase in reaction efficiency. That is, an increase in reaction rate. But in biological systems increase in reaction rate alone, if uncontrolled, is not necessarily advantageous. Just as important is the ability to regulate the reaction rate of a particular process. Although initially a variation in reaction rate could depend on substrate availability, ultimately the enzymatic complex responds to a variety of external inputs that together regulate the reaction rate. This is likely a major function of increased molecular complexity.

A good example of a relatively simple complex molecular assembly is the bacterial RNA polymerase. This enzyme consists of a core of five polypeptides, $\alpha_2\beta\beta'\omega$, required for the elongation of an RNA molecule and a suite of initiation factors (sigma factors) that associate individually with the core enzyme and direct the polymerase to the transcription start points of different classes of genes. Bacterial cells usually live in widely fluctuating environments where the availability of nutrients and oxygen can vary dramatically and so need to adapt the pattern of gene expression to changing conditions. Much of this adaptation is effected by altering gene transcription patterns and involves regulation by small molecules, substitution of sigma factors and changes in the ability to interact with other proteins regulating transcription. In a favourable environment some bacteria grow very rapidly, doubling in number every 20'. But should an essential nutrient, perhaps an amino acid, be exhausted the cells shut down the synthesis of major types of RNA, including the RNA molecules found in ribosomes. In the γ-Proteobacteria this shutdown is mediated by a small molecule, the nucleotide ppGpp aka 'magic spot', which binds directly to RNA polymerase and reduces the rate of initiation of these particular types of RNA. The principal subunit involved in this regulation is ω. Under more prolonged stressful conditions growth is shut down, and substitution of the sigma factor supporting rapid growth by another supporting only limited or no growth occurs. Again the synthesis of the most abundant classes of RNA is curtailed. ppGpp and sigma factor substitution are examples of mechanisms that regulate the overall patterns of transcription. But more specific mechanisms operate at the level of particular genes. Regulatory proteins – transcription factors – bind to specific DNA sequences in the vicinity of a transcription start site and can interact with RNA polymerase facilitating the initiation of transcription at particular genes. Different transcription factors such as these can interact with different polymerase subunits, frequently one of the α subunits or the σ factor required for rapid growth. These modes of regulation are largely concerned with starting the transcription of a particular gene, but the other stages of transcription – elongation and termination of an RNA molecule – are also regulated and also require direct interactions of other proteins with the transcribing enzyme. Because each interaction of RNA polymerase with other molecules, small or large, has specificity its structure must accommodate a substantial number of possible interactions. These are made possible by the structural complexity of having several different subunits, each of which can make different external contacts and so modulate different aspects of polymerase function.

Macromolecular complexity provides a means for different proteins to communicate with others – often many others – in a cell and at the same time can serve to modulate regulation by small molecules. Essentially it increases organisation and enables adaptation to changing external and internal conditions. But there is a cost. More proteins need to be synthesised and more encoded in the genome. When the need for adaptation is reduced, a corresponding reduction in complexity may be selected. One example of such a phenomenon is the bacterium *Buchnera*, which is found as an endosymbiont in specialised cells of aphids (Buchner, 1965). This intracellular environment is (presumably) stable relative to the fluctuations encountered by its close relative, the gut bacterium *E. coli* and importantly the nutrient supply would be more constant. For growth in this benign situation the *Buchnera* genome requires less information and indeed is seven times smaller than that of *E. coli*. But what are the crucial differences between the genomes? Compared with *E. coli*, the intracellular environment obviates the need for the adaptive stress response activated by drastic variations in oxygen and nutrient supply. One result is a reduction in the complexity of RNA polymerase. Instead of the seven different RNA polymerase sigma factors found in *E. coli*, *Buchnera* has only two – one required for growth and cell division and the other for a response to rapid temperature increases. Not only is the number of sigma factors reduced but also the proteins involved in regulating the substitution of one sigma factor for another. Similarly the RelA protein responsible for the synthesis of the stress alarmone ppGpp is absent in *Buchnera*. The number of different abundant DNA binding proteins that organise the structure of the bacterial chromosome is also reduced with those that are present in non-growing *E. coli* being absent. *Buchnera* is thus characterised by a substantially reduced organisational network that is compatible with the more limited adaptational responses appropriate for its lifestyle. At the molecular level there is less complexity, and the corresponding genomic reduction is a superb example of adaptive evolution.

Reductions in organisational complexity may also be induced by virus infection. Some DNA bacteriophages, instead of encoding their own RNA polymerase, subvert the host enzyme so that it becomes dedicated to the transcription of viral genes. For example, the phages T2 and T4 modify the structure of the α subunit of the core enzyme so that potential interactions with transcription factors are abrogated. The selective transcription of the major bacterial RNA species is also abolished. Finally, the main host sigma factor is inactivated, permitting its replacement by an alternative sigma factor responsible for directing the production of the mature phage particles. In this rather specialised case the network of interactions normally working in the host bacterium no longer works, and the effective complexity of the system is much reduced.

A particular lifestyle or previous evolutionary history may preclude further increases in the complexity of a particular macromolecular assembly. Why, for example, is the bacterial RNase P limited to 2 components while those of eukaryotes contain up to 10? Why have the different structures of the bacterial and eukaryotic RNA polymerases – and many other assemblies – become essentially fixed over long periods of evolution? Even now the organisation of the chloroplast polymerase is essentially the same as that of present-day bacteria. Perhaps the core function of the assembly has been optimised. But this would not preclude the evolution of more cellular complexity. The existence of a complex multisubunit molecular assembly could provide a means for recruiting additional components of a network at a distance. That is, relevant information about the state of a cell can be communicated from one complex to another through intermediates and not direct contact.

Whatever the mechanism, increasing complexity requires more information – information that is encoded in DNA. But this additional information can be derived from more than one source – not just from the expansion of a given genome but from the functional cooperation of two or more genomes in a biological system.

Cellular Complexity

Formally cellularisation can be regarded as an extension of a compartmentalised chemical system that is internally cohesive and self-organised by liquid–liquid phase separation. The difference is that whereas communication within a collection of different liquid phases is likely largely, although cannot be entirely, internal, cellularisation clearly distinguishes the totality of the internal phases (the pan-internal phase) from a single external phase – the environment. Provided a sufficient barrier exists between the pan-internal and external phases, the two phases can differ substantially in ionic composition, reactant concentration and even pH. In the present day the cellular barrier takes the form of a lipid bilayer membrane containing proteins that span the bilayer and are responsible for the selective and directional transport of ions and simple organic compounds. The initial creation of a sufficient barrier to permit escape was likely an incremental process (Mulkidjanian, Bychkov, Dibrova, Galperin, & Koonin, 2012; Dibrova, Galperin, Koonin, & Mulkidjanian, 2015) involving successive steps for increasing the selective discrimination of the transport of different ions and also in the efficient utilisation of a hydrogen ion gradient for ATP production (Lane, Allen, & Martin, 2010).

Once established as free living organisms, the primordial cells would diversify into many different manifestations largely retaining a unicellular mode of existence. Nevertheless, multicellularity in which rather than an agglomeration of a single cell type, different cell types are associated in a single organism has arisen separately many times during the course of evolutionary time. Current examples would include the intimate association of two cells – a spore cell and a 'mother' cell – in the *Bacillus* bacteria, the formation of multicellular fruiting bodies in the both the bacterium *Myxococcus xanthus* and a eukaryotic counterpart, the slime mould *Dictyostelium* and, of course, very prominently most plants and animals. In multicellular organisms communication between cells of different types becomes paramount. Intercellular signaling is differentiated, and communication between widely spatially separated parts of the same organism is effected by widely differing strategies. Both plants and animals employ liquid transmission via pipes – xylem and phloem vessels for plants and the blood circulation for animals. Animals also utilise an extensive neural network for rapid transmission of information. Additionally, to communicate between individuals of the same and often other species both plants and animals employ VOCs (volatile organic compounds, for plants aka green leaf volatiles), which can signal, amongst other things, both danger and sexual availability. The release of one such compound is enabled by a tobacco hornworm, the larva of a moth *(Manduca sexta)*, munching the leaves of a tobacco plant. The main target of this VOC is not the tobacco plant itself, or other tobacco plants, but a hemipterid bug, which is then attracted to prey on the larva (Allman & Baldwin, 2010).

The functional differentiation of one cell type from another establishes heterogeneity, a key measure of complexity. The number of different cell types in a given multicellular organism can then be directly related to the apparent complexity of its organisation. By this measure the bacteria, fungi, algae and most protists are relatively simple with an approximate number of cell types varying between 1 and 5 (Raff & Kaufmann, 1983). For most plants (bryophytes (mosses), ferns, gymnosperms and angiosperms (flowering plants), the number is between 20 and 40. In contrast in most animals the number of different cell types is >40 and for mammals can exceed 100. This variation conceals an important correlation. Within a given group – for example, plants and animals – the average number of cell types increases as the group evolves. In the case of plants, the evolution from mosses to ferns to gymnosperms and then to angiosperms is paralleled by successive increases in average cell type number. Similarly for animals in one evolutionary line average cell type number increases from echinoderms to tunicates (sea squirts) to teleosts (bony fishes) to amphibians and finally to mammals. The conclusion is that, on this parameter, complexity increases as evolution progresses. Not only does the average complexity increase but also there are cases where the diversity of a less complex group falls after the emergence of a more complex group. During plant evolution an increase in the diversity of gymnosperms was paralleled by a decrease in the diversity of ferns and subsequently gymnosperm diversity decreased as angiosperm diversity increased. Similar patterns occur in the evolution of animals (Niklas, 1986).

Nevertheless, although there is an apparent increase in complexity, there is no correlation between the number of cell types and the total DNA content of an organism (Raff & Kaufmann, 1983). This observation clearly questions the assertion that an increase in complexity is accompanied by an increase in information. But only if the sole source of information available to an organism is that encoded in its own DNA. This is unlikely to be the case. Organisms do not under natural conditions live in isolation but instead occupy a particular niche in an ecosystem comprising many organisms, each of which contains its own genome. In essence such a suite of organisms cooperate informationally (see Chapter 5 for a fuller discussion). In addition the information content of a genome is not necessarily directly related to its size. Many genomes, particularly those of higher plants, contain an abundance of apparently redundant multiple repeated sequences, some of which are highly mobile and were initially thought of as 'junk' DNA (see Chapter 6). But, perhaps more importantly, within the course of evolution not only has the processing of DNA information become more sophisticated but also organisms have developed new modes of information transfer, most obviously a nervous system.

The Evolution of Information Processing

Modern-day computing has seen an explosive evolution in the efficiency of information processing. From the mechanical underpinning of Babbage's analytical machine to the punched tape and then the punched cards of the 1950s to present-day high-speed chips and flash storage, the speed of information transfer has increased by many orders of magnitude (see Chapter 1 for a description of the state-of-the-art procedures in the early days of molecular biology).

In biological systems a fundamental aspect of information transfer between nucleic acids is the often transient association of complementary nucleic acid strands to form a double helix. These interactions may create a DNA–DNA, a DNA–RNA or an RNA–RNA duplex. Arguably during the course of biological evolution there has been a trend for the number and variety of these interactions to increase, particularly in the fine-tuning of the regulation of gene expression. In the RNA world, digital base-pairing within an RNA molecule could result in the production of a selection of secondary structures, and base-pairing between RNA molecules would be exemplified by codon–anticodon pairing. Base-paired secondary structures, either the looping of contiguous short sequences or the formation of a short stretch of double helix between more distance sequences define the overall tertiary structure of the molecule. Loops can influence the rate of transcription elongation as well as signalling termination and can also act as binding sites for regulatory proteins. Base-pairing between RNA molecules can also serve an important regulatory role. Short RNA molecules complementary to, say, a longer mRNA molecule can form a short duplex that can block the messenger function. For example, in *E. coli* the expression of two major membrane proteins, OmpC and OmpF, is controlled by the osmolarity of the external medium. When the cells are exposed to high osmolarity, a short RNA – micRNA aka mRNA-interfering complementary RNA – is transcribed from the opposite DNA strand immediately upstream of the OmpC (Mizuno, Chou, & Inouye, 1984). This short RNA can then pair imperfectly with the ribosome binding site on OmpF mRNA and so inhibit translation initiation (Figure 4.3).

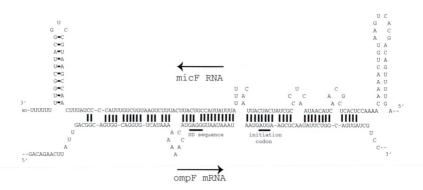

Figure 4.3 Base-pairing between a short interfering RNA (micF–RNA, upper) and the ribosome binding site, aka the Shine–Dalgarno sequence of the ompF mRNA (lower). Source: Reproduced from Mizuno et al. (1974), *Proceedings of the National Academy of Sciences, 81*, 1966–1970, with permission from Masayori Inouye

In addition to RNA–RNA interactions, RNA–DNA duplex formation can also be an important mediator of information transfer and plays a key role in the CRISPR bacterial immune system. This serves as a defence against bacterial DNA viruses. In a similar manner to the mammalian immune system the CRISPR locus contains a memory – copies of short DNA sequences derived from a previous virus infection. A subsequent infection then provokes transcription of the CRISPR locus. After processing into shorter RNA molecules, the appropriate RNA sequence can be matched with the corresponding DNA sequence in the invading virus via a DNA–RNA hybrid. This hybrid then serves as a guide directing the cleavage and inactivation of the viral DNA (Barrangou et al., 2007; Brouns et al., 2008). By providing a powerful means for editing DNA sequences, this seemingly otherwise obscure mechanism has proved to be of immense utility in scientific research (see Pennisi, 2013, for a description of the potential of the methods).

Another example of the evolution of information processing is the appearance of RNA splicing, a mechanism in which a section of an RNA molecule, usually an initial transcript, is precisely excised to yield a shorter molecule. Although examples are found in bacterial viruses, the phenomenon is largely restricted to eukaryotes and is more frequently utilised in evolutionary later forms. In some cases, for example, the *Tetrahymena* rRNA precursor (Cech, Zaug, & Grabowski, 1981), the excision can be regarded simply as a step in the process to yield a functional RNA molecule. In this context there is little contribution to complexity. However, for protein-coding genes the situation is very different. Splicing is principally a eukaryotic phenomenon, although examples are known in some bacterial DNA viruses and is often less frequent in organisms such as baker's yeast, with smaller genomes.

In higher eukaryotes many protein-coding genes have a modular structure (Box 2.5) in which different coding functional modules (exons) are separated by non-coding stretches (introns) in the DNA sequence. Transcription of the mRNA precursor includes both introns and exons. Splicing excises the introns producing a mature mRNA with the exons joined in coding phase. Because splicing can be selective in exon retention in the final mRNA product, the modular structure of the gene can be exploited to increase complexity. Instead of the gene encoding a unique translation product, selective splicing enables the production of different variants of a protein in different types of cell. These have similar functions but different amino acid sequences in some domains, while usually retaining the core exons essential to function. This arrangement permits a particular, say, enzyme activity to associate with different, but overlapping, suites of interacting molecules in different cells. In essence, a single gene now encodes, instead of one protein, several very similar proteins, thus increasing the available number of proteins without a corresponding increase in gene number. Selective splicing by adopting a novel mechanism of information processing can operate on systems on greater genetic complexity.

A modular organisation is also apparent in the DNA sequences regulating gene transcription – again especially in higher eukaryotes – where the control sequences specifying the temporal and spatial regulation in particular cell types can be separated into distinct blocks. Activation of a specified module then depends on the availability of the relevant transcription factor. This contrasts with the much simpler and shorter organisation in bacteria and in simpler eukaryotes such as yeast.

Genetic Programming and Epigenetics

What is epigenetics? The term was initially coined by Waddington to describe the study of how a single DNA genotype can instruct the production of a wide variety of cellular phenotypes during development. Subsequently it was widened to include meiotic or mitotic inheritance that did not involve a change in the DNA sequence. Common to these definitions is the notion of a cellular 'memory' that is maintained during cell division. However, it is now clear that many of the molecular mechanisms associated with epigenesis are particular examples of mechanisms involved in the maintenance of a certain state of the information-processing machinery. For example, the maintenance of an 'open' chromatin state competent for transcription is dependent on chemical modifications of the histone proteins. In principle this is little different from the selective silencing of different genes in different cell types during development. The same, or a similar, mechanism has been co-opted for different purposes. In this context, any requirement for cell division creates an arbitrary distinction, especially because the term *cellular memory* is just that – it does not by itself imply cell division as an integral part of the definition. The 'memory' in this case is an adaptation to the previous history of a cell and involves chemical changes in the information-processing machinery. To accommodate the new perspective, Bird (2007) proposed the most inclusive definition of epigenetic events as 'the structural adaptation of chromosomal regions so as to register, signal or perpetuate altered activity states'. These events act on cellular information flow and most likely comprise a collection of different subroutines whose effects are context dependent. They are important players in the execution of the DNA programme but do not by themselves change the overall set of instructions in the DNA sequence (Figure 2.10). Epigenetic marks such as 5-methyl cytosine are normally added during the life of an organism and are transmitted to a subsequent generation – they are an acquired characteristic with a 'deliciously Lamarckian flavour' (Bird, 2007) – see also Box 6.1.

Epigenetic events, especially those associated with transcriptional dynamics, can be transient, but some can also persist through multiple cell cycles. Commonly they involve the chemical modification of either the DNA itself or very frequently a histone protein. For example, a particular lysine residue, the N-terminal tail of histone H4, can be acetylated. Acetylation of this lysine can act as a 'tag', which can be recognised by a protein module – the bromodomain – found in many protein complexes that manipulate chromatin structure (Figure 4.4). Not only may an acetylated lysine act as a ligand identifying particular nucleosomes, but also lysine acetylation can loosen – albeit to a limited extent – the wrapping of DNA around the histone. Such a dual role is also characteristic of another histone modification – phosphorylation. The intrinsically disordered tails of linker histones that neutralise the negative charge on the linker DNA between nucleosomes frequently contain a short motif – SPKK or a closely related sequence – which can be modified by phosphorylation of the serine residue. By altering the charge balance between the linker DNA and the protein, this phosphorylation promotes a loosening of the structure of the chromatin fibre but in so doing reduces the probability of coacervate formation between the C-terminal tail of the histone and the linker DNA (Turner, 2018). In turn this effect has the potential to alter the compartmentalisation of chromatin function within the nucleus. This type of modification, although its effect is primarily physical, would be included in Bird's

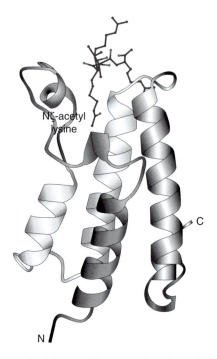

Figure 4.4 The bromodomain of the yeast histone acetyltransferase Gcn5 binding to its ligand N-acetyl lysine. *A black-and-white version of this figure will appear in some formats. For the colour version, refer to the plate section.* Source: Reproduced with permission from Owen et al. (2000), with permission from John Wiley & Sons, Ltd. Copyright 2000 European Molecular Biology Organization

definition. Another histone modification – the trimethylation of a particular lysine residue at position 9 in the N-terminal tail in the histone H3 – has a similar effect on compartmentalisation, although the mechanism is different. The modification is recognised by hetero-chromatin protein1, aka HP1, which associates with nucleosome arrays containing this modification thereby promoting chromatin condensa-tion and an associated liquid–liquid phase separation (Larson et al., 2017; Strom et al., 2017). The phase separation, in contrast to that of the linker histone, is facilitated by another modification, the phosphoryl-ation of HP1 itself.

Many of these histone modifications are not permanent. Not only are there enzymes in the nucleus, which add acetyl or methyl groups (or one of the many other modifications), but there are also enzymes that remove them. Depending on the relative rates of addition and removal the modifications are thus often in dynamic flux. Further unlike DNA replication, which is extremely accurate and equipped with error correction mechanisms, the accuracy of epigenetic tagging, although high, is significantly less (Bird, 2007) and is therefore less suited to the conservation of genetic information.

The other major type of modification found in eukaryotic nuclei is the methylation of cytosine residues to form 5-methyl cytosine. In vertebrates this occurs at the sequence CG and possibly as a consequence the frequency of occurrence of this nucleotide in vertebrate genomes is very much lower than would be expected from the genomic base composition. Cytosine methylation is associated with stable gene silencing. Its presence can either antagonise the binding of activating transcription factor or recruit a methyl-CpG-binding protein (MeCP2), which acts a transcriptional repressor or, in some contexts, as an activator. Cytosine methylation is largely responsible for parental imprinting where only one of the two parental alleles, usually the paternal one, is modified and consequently skews the parental contributions to gene expression during the development and subsequent growth of an organism. Plants differ from animals having the expected frequency of occurrence of the CG dinucleotide and exhibit an extensive, although variable extent, of cytosine methylation. In plants, although not in animals, this modification is most frequently associated with transposons, aka 'jumping genes' (see Chapter 6). Although cytosine methylation expands the repertoire of information processing, there is a cost. This methylation increases the mutation rate of the genomic sequence.

5

Cooperating Genomes

The result of the coadapting selection is a harmoniously integrated gene complex. The coaction of the genes may occur at many levels, that of the chromosome, nucleus, cell, tissue, organ, and whole organism.
—Ernst Mayr, 1963

The doctrine of the extended phenotype ultimately requires us to acknowledge the same kind of interactions among genes of different gene-pools, different phyla, different kingdoms.
—Richard Dawkins, 1982

Almost by definition, increases in biological complexity, whether at the molecular or cellular levels, or even at the level of species diversity, imply an increase in the effective information content of DNA in the system as a whole. An important question, which goes to the heart of the concept of biological complexity, is to ask to what extent any diversity-associated increases in information content are essentially discrete or are part of a continuum in which the evolution of complex biological systems is linked to incremental accretions of DNA information.

The recognition that different organisms of widely different phylogenetic backgrounds can coexist in a stable organisation and in so doing may create a completely different 'organism' has a long history (Martin, 1999). In 1867 a Swiss botanist reported that lichens are a 'consortium' of two organisms, a fungus and a photosynthesiser (Schwendener, 1867).

Subsequently, another botanist, Mereschkowsky (1905), following a tentative aside by Schimper (1883), made the then-radical suggestion that chloroplasts originated from endosymbiotic cyanobacteria (aka blue-green algae). Later Portier (1918) extended this idea to explain the origin of mitochondria. More recently Bonner (1988) propounded a modern foundation for these phenomena generalising in the concept of 'genomic integration' in which different DNA genomes, encoding different information, combine to specify a system. In this scenario the properties of the system in its entirety are determined not by a single DNA genome but by the interactions between the component genomes. The component genomes then comprise the system's 'hologenome' (Rosenberg, Koren, Reshef, Efrony, & Zilber-Rosenberg, 2007). In the simplest cases, as in some bacteria, there is only a single chromosome, but this is likely the exception rather than the rule. Many bacteria carry autonomously replicating plasmids, which, for example, can confer resistance to certain antibiotics. Here the phenotype of the organism is specified not by the bacterial genome alone but by the hologenome comprising the bacterial and plasmid chromosomes.

Examples of hologenomic complexity are ubiquitous and arguably underlie increases in biological complexity. A simple eukaryotic cell – such as a yeast or an animal cell – contains, in addition to the organism's nuclear genome, a separate DNA molecule in the mitochondria. This mitochondrial genome is thought to have been introduced by the symbiosis of an aerobic bacterium and a prototypic nucleated cell all, contained within the same membranous sac. Although in its current state the genic content of the mitochondrial DNA is probably much reduced from its original form with many essential genes being subsumed into the nuclear genome, the mitochondrial genome is still maintained as a separate entity encoding a few genes essential to mitochondrial function. Plant cells constitute yet another example of genomic integration. In addition to mitochondria, they also contain chloroplasts, essential for the capture of light energy by photosynthesis. Chloroplasts, which are likely descended from symbiotic blue-green algae, possess DNA molecules comparable in size to those of many bacteria. The hologenome of plant cells thus contains three elements: nuclear, mitochondrial and chloroplast. Chloroplast DNA encodes many genes whose products are necessary for information processing. For example, the multisubunit RNA polymerase of chloroplasts is structurally very similar to those of bacteria and blue-green algae. (In contrast that of mitochondria, now encoded by the nuclear genome, is structurally much simpler). Here again, although the original event (or most probably events) establishing the symbiotic relationship between blue-green algae and nucleated cells likely occurred around at least three billion years ago, the algal and nuclear genomes remain separate.

But why is the separation of individual genomes maintained? One possible reason is that this organisation allows the preservation of different regulatory mechanisms, otherwise different information-processing systems, whose loss, once they are established, would potentially be lethal. In present-day chloroplasts, changes in the super-coiling density of certain genes in the chloroplast DNA are closely integrated with the circadian light cycle, potentially providing a link between the availability of energy and transcriptional activity. In bacteria, energy availability is converted into high negative supercoil-ing density in part by the action of DNA gyrase. In plants, although this enzyme is encoded by the nuclear genome, it is specifically targeted to both chloroplasts and mitochondria and not to the nucleus. The combination of DNA gyrase and energy-dependent variation is superhelicity, a characteristic of the analogue mode of regulating information flow in bacteria. The unanswered question is whether the chloroplast and mitochondrial genomes have been retained to a greater or lesser extent to maintain a direct coupling between energy input and the energy stored by DNA superhelicity.

The examples of mitochondrial and chloroplast genomes contained within a single hologenome also illustrate an important principle of hologenomic complexity. The combination of genomes in a single functional environment – in these cases a cell – is synergistic and results in the acquisition of 'emergent' properties such that the resulting com-bination can respond differently to the external environment in which it finds itself and so explore different niches. Put another way, the func-tionality of the system is altered by an increase in the informational complexity of the DNA. Notably not only do higher cells contain more than one type of genome, but also these genomes differ in their mech-anisms of information processing.

Among further examples of the multifarious variations on genomic integration could be included that of a bacterium, *Buchnera*, living inside the cells of an aphid. Here the hologenome comprises the DNA content of the *Buchnera*, mitochondrial and nuclear genomes. However, the DNA genome of *Buchnera* is much reduced compared to that of related bacteria. All the genes regulating the growth transition from exponential aerobic growth to a slow-growing or stationary state appear to be absent, presumably an adaptation to a stable aerobic intracellular environment. Consequently by itself *Buchnera* is unlikely to survive in a changing environment, comparable, for example, to that experienced by the common gut bacterium, *Escherichia coli*. For viability, *Buchnera* requires the information contained in the hologenome of its own and the aphid genomes. The parallel with a prototypic mitochondrion is obvious. On the same theme the endosymbiotic zooxanthellae – strictly, photosynthesising dinoflagellates – are harboured by cells of corals, sea anemones and jellyfish, where they translocate the products of photosynthesis to the host and receive in return inorganic nutrients. It is to this symbiotic combination of genomes that the term *hologenome* was originally applied. Here the symbiosis is facultative, dissociation of the dinoflagellates from the coral cells can occur, resulting the well-described phenomenon of coral bleaching. Yet another example of an endosymbiotic relationship, and hence an expanded hologenome, conferring emergent properties in the association of the nitrogen-fixing Rhizobium bacteria with the roots of leguminous plants.

The most fascinating and instructive examples of genomic integration are those involving fungi. Fungi are generally denizens of dark and largely anoxic environments (there are, of course, some important exceptions, such as yeasts), and in many cases their brief forays into the light simply enable the distribution of spores. They lack both chloroplasts and, in some species, even mitochondria. Consequently, selection for active associations where the fungus benefits from energy dependent on light or oxygen harvested by another organism has occurred very frequently. Such associations can range from a benign mutual symbiosis to a completely parasitic existence. A very striking example of (almost) mutualism is provided by lichens (Figure 5.1). In these associations a fungal partner associates with one or more algal partners – usually a blue-green alga or a eukaryotic green alga – to create an organism with immensely different capabilities from that of its component parts. On average lichens contain hologenomes of two to three (fungal nuclear, mitochondrial and blue-green algal) – depending on whether the fungus contains mitochondria or not – or four to five

Figure 5.1 Complexity from genome sharing. A lichen (*Caloplaca saxicola*) growing on a stone fence post. *A black-and-white version of this figure will appear in some formats. For the colour version, refer to the plate section.* Source: Photograph by the author

(fungal nuclear, fungal mitochondrial, chloroplast, algal nuclear and algal mitochondrial) components. In some lichens the primary fungal partner may, in addition to an alga, commonly be associated with two, or even more, other fungi, usually yeasts, although whether such yeasts play a role in the symbiotic relationship remains to be established (Pennisi, 2016; Tuovinen et al., 2019). Individually the component organisms, particularly the algae, are usually viable. In this association, which may be symbiotic or contain elements of parasitism by the fungus, the alga provides energy from photosynthesis while the fungal thallus constitutes the morphological framework. The alga can also convert atmospheric nitrogen into bio-available nitrogen. However, the association enables the expression of a suite of fungal genes that enable morphological diversification to form the typical lichen thallus. In some species this robust growth is equipped to colonise extreme or nutrient poor environments – rocks in Antarctica, gravestones – largely unavailable to other organisms. Even in relatively benign situations lichens can be important colonisers – in Cambridgeshire thatched roofs are often initially carpeted by branched *Cladonia* lichens. Lichens also constitute a key component of developing soils such as the cryptobiotic proto-soils typical of the American midwestern deserts. In these soil crusts cyanobacteria are present both as free-living and lichenised forms providing a nutritional basis for further colonisation by bryophytes and vascular plants.

Lichens constitute an example of a biological system in which there are very intimate molecular contacts between the two component organisms yet together combine to form an association that itself can be identified as a distinct 'organism' in its own right. A new more complex 'organism' is created by the integration of the information in two or more genomes. The genomes of the two main components remain separate and can even be separately propagated, yet the fusion – the hologenomic phenotype – is a recognisable entity. In this case the distinction between what constitutes a distinct organism with a single nuclear genome and a particular biological system with a specified hologenome becomes fuzzy. The presence of intimate and highly spe-cific molecular contacts manifest in such a system can be regarded as an extension of the molecular complexity within a cell in which most proteins talk to other most other proteins either directly or at second or third hand. The increased information content in the hologenome – reducing informational entropy – is thus again directly coupled to reductions in Boltzmann entropy defined relative to free components.

Again lichens provide an instructive example of this phenomenon. Lichens produce a startling variety of 'secondary' chemical com-pounds, almost all of which are unique to lichens. These chemicals are not involved in the primary metabolism of lichens but instead are produced as chemical 'by-products' of the fungal component. Nevertheless, in some lichens these 'by-products' can constitute up to 5% of the total weight of the organism. In contrast, an isolated fungal thallus lacking its algal partner can normally produce only minute quantities of these compounds. The symbiosis of alga and fungus thus enable the up-regulation of lichen 'by-product' produc-tion. These compounds are quite varied in structure and function. Some within the lichen thallus serve as light screens, while others, by virtue of their noxious taste, are believed to act as deterrents for browsing herbivores. Yet others have a structural role helping to repel water, provide air spaces within the thallus and even possibly confer rigidity on the thallus. And some even act as antibiotics to inhibit the growth of other fungi, vascular plants and even other lichens. A visit to a local churchyard or graveyard with lichen-encrusted gravestones often reveals the intense competition between neighbouring lichens where the boundaries between the plants are demarcated by precise lines. Many of these compounds are clearly associated with adaptation to extreme and/or nutrient-poor environments, and the regulation of their production is a compelling example of integration and cooperation between separate genomes in the same organism.

The same fuzziness is also apparent in many other frequently encountered biological associations. Particularly noteworthy is the widespread collaboration of mycorrhizal fungi with the roots of many vascular plants. Such collaborations are especially crucial for trees where the mycorrhizal hyphae can form a very extensive underground network, dubbed a 'wood-wide net' (Macfarlane, 2019). Notably, this networking system can enable a physiological connection between different species of plants. For example, in a Canadian forest Simard observed that growth of Douglas fir saplings substantially benefited from a mycorrhizal association with paper birch saplings (Simard et al., 1997). In such a mycorrhizal association, the fungus colonises the host plant's roots either intracellularly or extracellularly. In both types of colonisation there are again intimate molecular contacts, the major difference being that in intracellular colonisation the fungus penetrates the root cell wall, although not the outer cell membrane immediately underlying the wall. This symbiosis benefits the fungus by providing direct access to the simple carbohydrates, such as glucose, in the host plant. Concomitantly the host plant benefits by gaining access to the very extensive mycelial threads of the fungus. These threads are generally thinner than roots and consequently have a substantially greater surface:volume ratio. This increases the absorptive capacity for water and minerals that are then available to the host plant. The chemistry of absorption also differs between the fungal mycelium and the host roots enabling, for example, the host to absorb phosphates from basic soils where they might otherwise be unavailable. From this perspective a luxuriant beech hanger on a chalk hill is the phenotype of the hologenome comprising beech and mycorrhizal fungi partners. Associations of this type are probably evolutionarily ancient. The approximate times of the origin of the endomycorrhizal fungal lines are closely contemporary with the colonisation of land by ancient plants.

Closer to home an individual human does not exist wholly independently. Instead, the human body is home to an array of commensal microorganisms, including many varieties of bacteria fungi. Many of these organisms are true commensals – that is, they live, for example, on the skin scavenging dead sloughed cells – a simple ecological niche. Others, particularly those in the gut, play a more active role, participating in the digestion of certain molecules in foods and also again making direct and specific molecular contacts with cells lining the villi of the small intestine. The precise composition of the gut flora varies substantially between individuals and to a certain extent determines dietary responses. In humans the essential nature of the gut flora may be less immediately apparent than, say, in cows, where the bacteria in the rumen are necessary and crucial for the digestion of cellulose. Here the phenotype of the total hologenome of the cow, comprising the cow nuclear genome and the genomes of all the associated bacterial and fungal genomes, is essential for the existence of what we recognise as a cow. Without the bacterial cellulose fermenters in the rumen, there would be no cattle as we know them.

Dawkins in 'The Ancestor's Tale' – an entertaining scamper back along the animal evolutionary tree – provides a superb and more extreme example of a similar association. Termites, like cows, digest cellulose. In their case the cellulose is from wood, in cows from grass. Whereas some termites – 'advanced' – can manufacture sufficient of their own cellulase to digest cellulose, others rely on their gut microbiota to provide the enzymatic activities for efficient digestion. One component of this microbiota is a motile protozoan *Mixotricha paradoxa* (Figure 5.2). This organism is covered with tiny 'hairs', initially described as cilia because they resemble a specific type of hair-like structure on the surface of animal cell and of another group of unicellular organisms appropriately termed Ciliates. But the 'cilia' coating *Mixotricha* are not true cilia. They are actually tethered spirochaete bacteria, both long and short. Bacteria in this group normally swim freely with an undulating motion. On *Mixotricha*, however, they

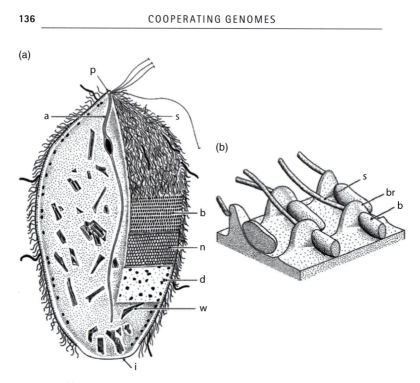

Figure 5.2 (a) The protozoan Mixotricha paradoxa. The organism is drawn in the optical section on the left, while on the right the surface structures are shown at a series of progressively deeper levels passing backwards. a, axostyle; b, bacterium; d, dicytosome; i, ingestive zone; n, fibrous network; p, papilla; s, spirochaete; w, wood. (b) Reconstruction of a small area of the cell surface to show the shape of the brackets (br) and the arrangements of the bacteria (b) and spirochaetes (s). Source: Reproduced with permission from Cleveland and Grimstone, 1964

undulate in unison and are fixed on little platforms formed by a regular array of 'brackets' on the surface of the protozoan. These brackets align the spirochaetes so that they all point in one direction. However, real cilia not only comprise a hair-like structure, but also have a 'basal body' at their root. Spirochaetes, presumably because they swim freely, lack this appendage. Remarkably, though, the spirochaete cilia mimics in *Mixotricha* are also associated with a basal body–like structure. But again this is yet another different organism – a pill-shaped bacterium. So the *Mixotricha paradoxa* that we observe comprises (at least) three cooperating principal genomes – those of the protozoan itself and those of the associated spirochaete and pill-shaped bacteria. In Dawkins' words, 'It becomes quite tricky to distinguish between "own" and "alien" bodies in such cases'. The genetic information is integrated to the extent that the *Mixotricha* genome is involved in the production of the regular surface brackets essential for the directionality and coordination of the spirochaete association. This mode of genetic integration is far from unique. Other protozoans in the termite gut also acquire motility by association with different bacteria but at least some of these cases the motility is not provided by the undulations of a spirochaete but instead by a different structure, a whip-like flagellum, which again imparts motility to bacteria.

Intimate associations of the type demonstrated by *Mixotricha* and its 'friends' raise the question of how much the expression of the associated genomes is coordinated. So far little is known about this, but both starvation and antibiotics, which target the bacterial components, produce dramatic changes in the morphology of the constituent organisms. Both environmental insults result in the disintegration of the molecular attachments between the spirochetes and the protozoan. *Mixotricha* becomes spherical and immotile, while the spirochaetes transform to small spherules or cysts. While production of these dormant spherical forms may be a survival mechanism in response to environmental stress, viability declines. Here the symbiosis or genetic integration is important for survival.

The genomic integration, of course, does not end simply with the association of *Mixotricha* and its bacterial commensals. The utilisation of cellulose as an energy source by 'lower' termites is entirely dependent on their gut microbiota. While even lower termites secrete a cellulase from their salivary glands and their midgut, the presence of gut protozoans is essential for termite survival on a wood diet. These protozoans, of which *Mixotricha* is but one example of many, provide cellulolytic enzymatic activities that enable the efficient breakdown of the lignocellulose complex in wood fibres. In addition to their role of providing motility to certain members of the protozoan community, some of the bacterial associates, including some that adopt an intracellular location resembling that of *Buchnera,* fix nitrogen, potentially providing an essential nutritional supplement to an exclusive diet of wood. The overall number of organisms existing in the termite gut is likely huge and part of an extended symbiosis. A description of the complexity of this minute biological niche would be correspondingly enormous, perhaps mirroring in miniature the complexity of the biological world as a whole. The corresponding hologenome, comprising all the disparate information in the component genomes encoding multifarious enzymatic functions, would again be probably orders of magnitude larger than a single nuclear genome. While this degree of complexity is (or should be) awe-inspiring, it is worth repeating that it is only a very small part of the entire biological system, perhaps comparable to a planet or solar system in the astronomical universe.

Not all the individual organisms comprising a hologenome as discussed earlier are necessarily benign to an individual organism. Pathogenic gut bacteria – of which *Salmonella* is a good example – produce proteins that form direct contacts with others lining the villi. These contacts then facilitate the invasion of the lining by bacteria with often severe consequences for the host organism.

But where to draw a line between an organism with its own nuclear genome and an ecosystem? Both mycorrhizal associations and human beings could be and have been described as ecosystems in which the component organisms are capable of living without the others – although the quality of life of a human without gut bacteria is likely significantly diminished. Similarly, vascular plants without their mycorrhizal partners lack vigour unless they have an adequate supply of externally applied nutrients. Again, *Mixotricha* and its friends are mutually dependent on each other for viability. But do these associations constitute an ecosystem, and if not, what are they? For lichens without the symbiotic alga or blue-green alga, the fungal component can often survive but in a less robust state, although the algal partner can often, and does, exist independently. For the symbiosis of coral cells and zooxanthellae the same situation applies. Without the zooxanthellae, the corals become more vulnerable. The difference between lichens on the one hand and mycorrhizal associations on the other is not absolute but one of degree. The common feature is that the properties of the system as a whole are dependent on the information contained in the DNA of the relevant hologenome.

The examples discussed so far all involve intimate molecular contacts between what might normally be considered distinct organisms. But the availability of a free-living phase of the participating components in both lichens and corals raises the question of whether only the independent free-living phases can properly be described as components of an ecosystem. But if not, how to describe possibly transitional situations where, for example, an organism like *Buchnera* has taken up residence inside a cell of another organism but has become completely dependent on some of the services provided by the host, and may in return provide a useful nutritional supplement. In biology the variety of possible and actual associations is so large that any functional distinction between, on the one hand, the symbiosis of a mitochondrion and a nuclear genome, and on the other, an ecosystem in the normal understanding of the word seems arbitrary. There is no obvious distinct separation, and biological diversity ensures that whatever line is drawn there will always be exceptions. In this context Bonner (1988) expanded his concept of genomic integration to include the transient behavioural associations of cleaner fish with usually much larger hosts. These small fish maintain so-called cleaning stations, where other fish congregate and perform specific movements to attract the attention of the cleaner fish. Remarkably, these small cleaner fish will safely clean large predatory fish that would otherwise eat small fish such as these. In principle the cleaner fish differ little in function from the bacteria and other organisms living off dead human skin. In both cases, as in any symbiosis, the interaction is beneficial to both participating parties. The well-being of the both the cleaner fish and the larger fish thus depends on the genomic information encoded by both organisms, and the simple system can be described by the hologenome specifying both entities.

But other organisms also interact with either cleaner fish or their hosts. Again the specification of the system described depends on the number of individual genomes involved. Importantly as the number of individual genomes increases, so does the overall information content. This extends the hologenome concept from binary interacting organisms essentially to whole ecosystems, where the interactions between the component organisms form a network. When applied to a general type of habitat, such as a rainforest, the term *ecosystem* is well understood, but the organismal components that comprise an ecosystem often vary both temporally and in location reflecting small variation in environmental conditions. The networks are in flux and so too are the hologenomes. Again one can make an analogy with a single nuclear genome. Here the genomic DNA sequence itself is in flux – it is changed by mutation virtually every generation. For example, genomic DNA sequencing reveals that on average a human nuclear genome accumulates estimated changes equivalent to ~1.1×10^{-8} per base-pair per generation (Roach et al., 2010). This alters the available pool of allelic variants. In the same way, in rainforests, species of a particular type of organism often have a very limited range but may live in close proximity to very similar closely related species existing in a marginally different milieu (where the differences may simply reflect a different suite of the same spectrum of organisms or slightly different environmental conditions). On this argument the component genomes of a hologenome are equivalent to genes of a single genome, with very closely related species, for example, Darwin's finches (the Geospizinae of the Galápagos), constituting the equivalent of allelic variants of genes.

Considered in isolation, Darwin's finches are a paradigm of the evolution of different forms from a founder population. While the genomes of individual species probably differ little in information content (although it would be instructive to find out to what extent this is so), the specific adaptations allow different species to inhabit different environmental niches and hence contribute to distinct hologenomes, which are able to then diverge and increase overall diversity. This process is thus distinct from the co-adaptation in symbiotic or more generally interdependent, such as predator and prey, biological associations.

Discussions of the direction of biological evolution – with the prescient exception of scientists like Lynn Margulis – are often framed in the context of organisms, very often just animals. But it is the divide between eukaryotes and bacteria that constitutes one of the fundamental roots of evolutionary divergence and is possibly the most spectacular and most far-reaching example of the creation of a hologenome involving an intimate extension of genomic integration. But this same process is also responsible for other important increases in biological diversity – the creation of plants, including red algae, by the incorporation of a cyanobacterial genome into a eukaryotic cell and of brown algae by the inferred fusion of two eukaryotic cells, one of which contained a plastid. Within these distinct categories of hologenome the DNA genomes remain discrete. Gene movement may occur, usually by transfer to the nuclear genome, but the over long periods of evolutionary time since the establishment of these categories the different DNA genomes associated with chloroplasts and mitochondria have not been assimilated into a single unitary genome.

Although mitochondrial and chloroplast genomes have been conserved, even if to varying extents, in some organisms there are other intracellular organelles, apparently closely related to mitochondria, which lack DNA. These organelles, like mitochondria, are bounded by a double membrane and contain proteins that are closely evolutionarily related to certain mitochondrial proteins. In both mitochondria and chloroplasts the production of ATP is mediated by the passage of electrons along a chain of electron donors and acceptors. In mitochondria the final electron acceptor is oxygen, which is reduced to water. However, many organisms live in environments that lack oxygen. These organisms may still contain mitochondria, but instead of oxygen, they use organic or environmental compounds such as fumarate or nitrate as the final electron acceptor. The organelles, which lack DNA but still appear to be related to mitochondria, include hydrogenosomes, which produce hydrogen by the reduction of a proton and concomitantly produce ATP but lack an electron transport chain, and mitosomes, which do not produce ATP and again lack an electron transport chain. The retention of a DNA genome is associated with the presence of a core component – so-called complex I – of the electron transport chain. Whether this is essential for the preservation of the organelle genome is an open question, but loss of certain enzymatic functions of mitochondria may eventually result in the total loss of an independent DNA genome.

Evolutionary Energetics

It is axiomatic that living forms require a source of energy for growth and reproduction. This source can be chemical, comprising simple compounds, which are then metabolised. Of these sugars are a notable example. However, the dominant current biological energy source is sunlight, which drives the conversion of water to oxygen in chloroplasts, a reaction coupled to the production of low-molecular-weight substrates for metabolism.

Energy availability is often equated to the input from a particular source, for example, the sun. While the overall amount of energy available is obviously crucial for the functioning of an ecosystem, another important parameter is the efficiency with which the available energy is utilised. (This is a property of any ecosystem using the broadest understanding of the word. It is as equally applicable to biological ecosystems such as a rainforest or coral reef as it is to man-made complexity, i.e., ecosystems such as a city or an industrial complex.) In biological ecosystems energy efficiency is likely related to the complexity or heterogeneity of the system. This is simply a consequence of the utilisation of light, the primary physical energy source, to create disparate biological materials that can then be recycled to serve as secondary chemical energy sources. Depending on the complexity of the system – the overall number of components – this recycling process can, by increasing efficiency, both conserve essential nutrients and, within defined limits, increase the overall stability of the complete chemical system defining the ecosystem.

On the long timescale of biological evolution, energy availability itself evolved, sometimes by an emergent leap and other times by a more incremental process. Emergent leaps would include the generation of ATP by a hydrogen ion gradient across a barrier (chemiosmosis) (Chapter 7), the adoption of photosensitive molecules such as chlorophyll to harvest light energy and the integration of proto-bacteria and cyanobacteria into eukaryotic cells as mitochondria and chloroplasts (this chapter). Of these, the adoption of photosynthesis had profound consequences not only for energy availability per se but also for the influence of its principal by-product, oxygen, on the evolutionary trajectory of the planet. On a more modest scale, one example is an increasing complexity of the light harvesting system provided by addition of antennae enabling the feeding of photons to chlorophyll. This resulted in improvements in the efficiency of photon capture and thus of energy utilisation. It is likely this evolution of energy availability that underpins the rise of the information capacity of biological systems.

6

DNA, Information and Complexity

The evolution of biological systems as we understand them has been accompanied by a general increase in the complexity of their organisation. Comparing more complex organisms to simpler ones, this is apparent at many levels – a substantially increased amount of DNA, more polypeptides are required involved in the same task, a strong tendency to multicellularity, an increased sophistication of secondary information systems such as epigenetics and pheromones, mutual dependence and cooperation between disparate genomes, and of course, the evolution of a centralised information-processing device – the brain. Such progressions raise the issue as to whether the all increases in complexity are ultimately driven only by a DNA genome both containing more information and utilising this information more efficiently, or is such a perspective a reversal of the actual situation? Put simply, are there other, additional, sources of information that act as drivers of biological evolution and complexity? In other words, is it primarily an increase in total effective information content rather than just that in DNA that is responsible for the complex biological world in which we exist? If so, then a dominant role of DNA in specifying a system on which natural selection acts comes into question.

The Nature of Complexity

The complexity of a system, as loosely understood in this context, can be regarded as a measure of the dynamic interconnections between its disparate components. Increases in the interconnections – and hence the organisation of such a system – can ultimately be expressed thermodynamically as a decrease in the intrinsic entropy of the interconnected entity. This essential property implies that not only is energy required to promote the evolution of complexity, but also that more complex systems require a greater energy input for their maintenance – unless they evolve more efficient mechanisms for energy utilisation. In biology the connection between energy and complexity is well illustrated by the obligate symbioses between the mitochondrial and nuclear genomes in most eukaryotic cells, between the chloroplast (cyanobacterial) and nuclear genomes in algae and plants. Similarly, symbiosis between the apicoplast (derived again ultimately from cyanobacteria) and nuclear genomes in certain protozoan parasites such as Plasmodium may originally have enhanced energy availability. And again, some coelenterates harbour intracellular photosynthetic symbionts – green algae in the case of Hydra and zooxanthellae (a type of dinoflagellate) in coral polyps.

In addition to this energy requirement, there are also other costs associated with complexity. These can include specialisation and a restriction to survival within prescribed environmental limits. For many multicellular organisms where diversity of cell type is accompanied by genetic diversity, senescence may be another consequence. The possible dangers of specialisation are well illustrated by the example of the fig wasp. Fig trees have developed a mutualistic association with a type of wasp that is essential for fig pollination. Conversely, the fig flower is essential for wasp reproduction. This system is maintained by adaptation and co-evolution resulting in different fig species being pollinated by different species of wasp. A wrinkle in this system is that other wasps parasitise the pollinating fig wasps, which thus become essential for the reproduction of the parasites. Like any other interconnected but isolated system, the survival of the individual components depends ultimately on the survival of all of them.

The phenomenon of cooperating genomes is pervasive in biology and is essentially the antithesis of selection solely by competition. In a eukaryotic cell the nuclear and mitochondrial genomes do not compete with each other because they are both required for the viability of the system – in this example the cell – as a whole. There may be a gradual transfer of genetic material from mitochondrial genomes and a loss of non-essential genes in both, but as long as the energy producing electron-transport function of mitochondria is required, the mitochondrial genome is retained. Of course, there may be well competition between different complex systems – between different animals and plants for example – but genetic cooperation is, perhaps paradoxically, both usually an essential requirement and an enabler of such competition.

Adaptation and DNA Information

Although adaptive processes that result in the dynamic evolution of the fig–wasp partnership are ubiquitous and well understood, it is pertinent to ask to what extent the evolution of information processing at the genomic level aids, and may be essential to, adaptive processes. Such considerations are especially relevant in the mechanisms underlying the evolution of complex organisms. Adaptation of genomes in environments in flux requires the generation of sufficient genetic variation for natural selection to act. The evolution of genetic mechanisms, and so the DNA organisation, that facilitate the production of variation are crucial. One such major step was likely the emergence of linear chromosomes with protected ends – telomeres. This innovation allowed the reassortment of genetic material between the two copies of same chromosome in a diploid nucleus. Because the two chromosomes carry different allelic variants this meiotic reassortment increases the variety of possible allelic variations in the haploid germ lines and therefore the adaptive possibilities in any resultant diploid progeny.

Increasing the range of adaptive possibilities is not the only consequence of linearity. Arguably linearity may also promote biological complexity. In meiosis chromosome pairing occurs at multiple sites along their length. With closed circular chromosomes – typical of Bacteria and Archaea – multiple pairing, and the resultant intertwining of two circles, would create a major topological problem. With telomeres chromosomal identity becomes distinct and is associated with both localisation and chromosomal multiplicity within the nucleus. A further likely consequence of chromosomal linearity is linked to the possibility of extension, so potentially increasing the number of genes, and hence information, contained within a genome. The initiation of DNA replication in Bacterial and Archaeal circular chromosomes generally occurs at a single location – the origin – and then proceeds bidirectionally around the molecule until both replicating complexes reach the termination site situated directly opposite to the origin (Figure 6.1). Because the maximum rate of DNA synthesis is effectively fixed, a single origin restricts the size of a circular chromosome. The more DNA the chromosome has, the longer it will take to be replicated. For rapidly growing organisms like Bacteria, this is potentially a problem. The time needed to replicate the chromosome may exceed the doubling time of the organism. For a bacterium such as *Escherichia coli*, the doubling time is 20′, but the time required to replicate a whole chromosome is 40′. The imbalance is circumvented by the single origin firing twice during a generation. However, because the chromosome is a closed circle, multiple firing of a single origin is likely to increase the probability of DNA tangling within the chromosome. In contrast to Bacterial and Archaeal chromosomes, the linear chromosome of eukaryotes contain multiple origins of replication, which can be simultaneously active. The potential tangling problem is eliminated. Nevertheless, in

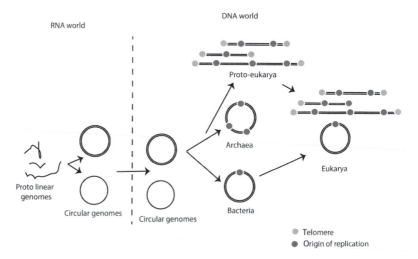

Figure 6.1 Evolution of chromosome organisation from proto-linear genomes in the RNA world to the eukaryotic chromosomes with telomeres and multiple origins of replications. A eukaryotic cell contains both linear and circular (in mitochondria and, in plants, chloroplasts) chromosomes. Bacteria also include cyanobacteria.

simple eukaryotes, such as budding and fission yeasts – the multiple individual chromosomes are of comparable length to circular bacterial chromosomes. It is only in higher eukaryotes (Raff & Kaufmann, 1983) that the expansion of the genomic DNA content is accompanied by substantial increases in both the length of the chromosomes and the gene content.

The increase in chromosome length correlates not only with increases in organismal complexity, but also with different routes to the generation of complexity. For example, plants and animals, both complex organisms, have evolved different modes of DNA information handling. Both have evolved into a panoply of multicellular variants, and both contain any different cell types. Typically the number of distinguishable cell types in animals is greater than that in plants. For example, in mammals there are >100 cell types, while in higher plants such as angiosperms the number is around 40 (Raff & Kaufmann, 1983). Although such estimates of internal diversity are perhaps highly subjective and may even be misleading, taken at their face value they imply by that plants are on average less complex than animals by this measure. Yet, with the notable exception of salamanders, the average angiosperm genome contains 5–10 times as much DNA as the average animal genome. If DNA content is taken as a measure of information content, these comparisons of the differences in complexity and genome size are counter-intuitive. A possible resolution of this apparent paradox lies in differences in the biological handling of genetic information and in the nature of the genomic organisation in plants and animals.

One notable difference between plants and animals in the handling is the timing of the segregation of the germ line. Whereas in animals the germ line is separated from the somatic cells at a very early stage in development – sometimes at the four-cell stage – in plants this early segregation does not take place, and the germ cells are formed from the apical shoots. In the germ line the genetic integrity of the genome is protected. Thus the early segregation in animals in principle could allow a greater internal expansion of genetic diversity as an organism matures than would be the case in plants. For them, totipotency must be preserved at multiple sites consequently limiting any internal expansion of genetic diversity. Indeed, in both plants and animals multicellularity is accompanied by internal genetic diversity. No longer are all cells restricted to a diploid number of chromosomes.

Many contain multiple copies – they are polyploid. Polyploidy is often associated with both larger cells and the production of large amounts of selected suites of proteins. Examples include muscle and liver cells in mammals and salivary gland cells in Dipteran insects. An extreme example is the 10^5- to 10^6-fold increase in chromosome ploidy in the glands producing fibroin – silk – in the moth *Bombyx mori*, in effect generating silk-producing factories. Similarly variations in the copy number of particular genes within normal chromosomes are widespread and contribute to many different phenotypes. Polyploidy is not the only type of cell-specific variation in DNA organisation. In the mammalian immune system the successive maturation of antibodies is accompanied by the generation of small DNA deletions. The resulting genetic diversity within a multicellular organism produces many different types of cells with a great variety of functions. These cells essentially cooperate in a highly complex system. But this complexity has a cost. Many, possibly most, of the cell types will have lost the ability to function as germ cells, either because they are no longer genetically identical to the early embryonic stem cells or because their environment precludes reversion to totipotency. Maybe one cost of multicellular complexity is senescence.

The difference in the developmental timing of the establishment of germ line between plants and animals implies that these groups have adopted different strategies to the evolution of multicellular complexity. To what extent is this reflected in their genetic material? On average, plant genomes contain substantially more DNA than animals. This does not reflect a significant difference in the number of genes but more a greater prevalence of short repetitive DNA sequences in plants relative to animals. The function of these sequences has long been controversial but is likely central to evolutionary mechanisms. For example, a pervasive competition model attains its apogee in discussions of this so-called junk DNA. In both plants and animals, coding genes account for a relatively small fraction of their genomes. Often the remaining sequences contain, amongst others, multiple copies of repetitive DNA sequences, which can have their origin in mobile genetic elements. These, normally short, DNA sequences, as their name suggests, can move from one part to another of the genome in which they are contained. The frequency with which this happens is quite variable and depends on the nature of the particular element involved. In the absence of any possible obvious advantage to the 'host' genome, these sequences were initially termed 'junk DNA' but latterly came to be regarded as the 'ultimate parasite' (Dawkins, 1976; Orgel & Crick, 198) exemplifying 'selfish' DNA. Yet, very early observations had shown that the great bulk of the eukaryotic genome is transcribed, albeit sometimes at low levels (Church & McCarthy, 1967). These RNA transcripts of course included RNA copies of junk DNA and imply that a cell may also produce 'junk RNA'. The term *ultimate parasite* implies that these DNA sequences confer no advantage whatsoever on the survival of remainder of the genome but are retained and replicated at its expense. The corollary of this argument is simply that if a junk DNA sequence could be shown to confer an evolutionary advantage to the genome in which it resides, it could no longer be thought of as an 'ultimate parasite' but rather as its polar opposite, an 'ultimate symbiont', or perhaps even something in between.

Nevertheless a clue to a possible cooperative and mutually beneficial function of mobile genetic elements comes from flowering plants. Many gardeners will have planted for decoration examples of plants with variegated leaves or even flowers. The vegetatively propagated varieties originally arose as 'sports' on their progenitors and are examples of somatic mutations. A fine example of such variegation is the patterned flower of *Rosa gallica versicolor* derived from *Rosa gallica*, the Rose de Provins (Figure 6.2). This botanical sport is associated in legend, probably mistakenly, with Fair Rosamund, a favoured mistress of Henry II. However, after some time – usually a few years – gardeners discover to their chagrin that some shoots lose the variegated pattern of leaves or flowers and revert to the original non-variegated form. Not only is the variegated pattern lost, but also the reverted shoots often grow more strongly, probably because non-variegated leaves contain more chlorophyll than their variegated counterparts. Then without the intervention of the gardener, the reversions simply out-compete the more delicate variegated forms.

Switches in flower colour are similar in character to switches in patterns of variegation and can occur with comparable frequencies. Although the molecular mechanism of switches of this type has been established for a few cases, for example, flower colour in the morning glory *Ipomoea*, the mechanism is likely more general and is an excellent example of DNA flux. The original somatic mutation and resultant phenotypic change result from the insertion, or transposition, of a piece

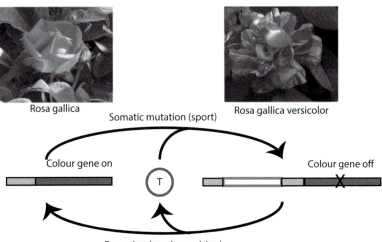

Figure 6.2 Variegated patterns in plant flowers and leaves can be due to the insertion of a short DNA transposon (T) in the control region (green) of a gene. When this DNA jumps out (precise excision), the original DNA organisation and the phenotype is restored. The different variants of *Rosa gallica* shown arose from the same rootstock in the author's garden.
A black-and-white version of this figure will appear in some formats. For the colour version, refer to the plate section.

of DNA – a mobile genetic element, aka 'jumping' DNA – into a gene determining flower colour or chloroplast function in a leaf (Figure 6.2). This event alters the expression of the targeted gene and may even completely shut it down. Usually, however, the insertion is relatively unstable and in a second event jumps out, precisely restoring the original DNA sequence and function of the gene. The excised DNA is then available for reinsertion elsewhere in the genome.

Transposition of mobile genetic elements is associated with an epigenetic tag – cytosine methylation – but is not the only type of epigenetic mechanism affecting flower form or colour. A naturally occurring mutant of toadflax (*Linaria vulgaris*) (Linnaeus, 1749) is one in which the fundamental symmetry of the flower is changed from bilateral to radial. The mutation affects the activity of the *Lcyc* gene, which in the mutant is extensively methylated and so silenced (Coen, 2001). Occasionally and at low frequency the mutant reverts phenotypically, correlating with demethylation of the gene and restoration of transcriptional activity. This 'epimutation' does not involve a change in the DNA sequence of the *Lcyc*, but like mobile genetic elements, the reversion of the phenotype is a somatic event and is subsequently heritable.

The phenomenon of genetic flux in this context has profound evolutionary consequences. A somatic mutation is expressed in a shoot and results in a changed phenotype that competes with other unmutated shoots. Crucially all the shoots have the potential to develop into flowers producing seeds and pollen. In other words, the somatic mutation, acquired during the growth of the plant, can be incorporated into the germ line and passed on. This is the essence of the theory of evolution proposed by the French evolutionary biologist Jean-Baptiste Lamarck (Boxes 1.1 and 6.1). Nevertheless, although somatic mutations of this type may be incorporated into the germ line in plants, they will still be subject to the normal assortment and segregation during meiosis. This situation points to a major difference in evolutionary strategies between plants and animals. Because many angiosperms have multiple flowering shoots, every plant then has in principle the potential to incorporate different, and possibly multiple, somatic mutations in each shoot. Such a strategy would increase – even if at a relatively low frequency – adaptive possibilities by increasing the variety of alleles from each plant available for selection. In this respect the difference between plants and animals is that whereas for animals inherited mutations are essentially those that arise in the germ line, in plants inherited mutations can arise both in the somatic tissue (although not in all locations) and the germ line. For a sessile organism such a plant surrounded by a variety of potential habitats, perhaps differing only slightly, the advantage is obvious. For more mobile animals this consideration would be less important.

Box 6.1 Lamarckism and Darwinism

While rightfully acknowledged as the originator of the concept of evolution, Lamarck's advocacy of evolution being driven by phenotypic characteristics acquired and expressed during an organism's lifetime has been largely derided and discarded by many evolutionary biologists. (To be realistic, the field in general, although with one or two commendable exceptions, is not noted for measured debate – indeed, of Lamarck's hypothesis one stated disparagingly that it was a 'universally admitted fact that the "Lamarckism" theory is false' [Dawkins, 1976]). In contrast, Darwin's monumental treatise 'On the origin of species by means of natural selection, or the preservation of favoured races in the struggle for life' was concerned mainly with the mechanisms by which assemblages of species evolved. He spelt out the principle of natural selection or competition between evolving forms. In this he was greatly influenced by the eminent Swiss botanist Augustin de Candolle, who had written of 'Nature's wars' or competition between organisms sometime previously. Indeed, in an essay written in 1844, Darwin acknowledged how strongly de Candolle had influenced his thoughts on natural selection. Introducing the topic he wrote (Darwin, 1842),

> De Candolle, in an eloquent passage, has declared that all nature is at war, one organism with another, or with external nature. Seeing the contented face of nature, this may at first be well doubted; but reflection will inevitably prove it is too true. The war, however, is not constant, but only recurrent in a slight degree at short periods and more severely at occasional more distant periods; and hence its effects are easily overlooked. It is the doctrine of Malthus applied in most cases with ten-fold force.

In this context Darwinism refers to the mechanism of natural selection that applies to the fitness of mutations. In Lamarck's model these mutations would be somatic, but as normally understood, often under the erroneous rubric of Darwinism, the mutations on which natural selection acts arise solely in the germ line. Although Lamarck himself trained as a botanist, most current protagonists in evolutionary debates are zoologists. And animals of course hive off their germ lines very early in development, and consequently mutations and subsequent reassortment arise virtually solely in the germ line. In plants, however, the germ lines are differentiated from, usually multiple, apical shoots at a much later stage, and consequently their germ cells can incorporate any somatic mutations generated and expressed during the growth of the shoot. Lamarck's tragedy was not the validity of his fundamental idea but his choice of an animal, the giraffe, as an illustrative example of an acquired characteristic.

If indeed certain mobile genetic elements can confer a selective advantage on the genome in which they reside, they would qualify as an 'ultimate symbiont' and not an 'ultimate parasite'. Recent studies have also indicated regulatory functionality for the transcriptional products of other DNA sequences of previously unknown roles. However, it remains a possibility that some DNA sequences, subsumed under the general definition of 'junk DNA' have no current function. They may, for example, be relics of genomic flux, and their residence may also be only transient.

7

Origins of Complexity

... if (& oh what a big if) we could conceive in some warm little pond
with all sorts of ammonia & phosphoric salts ... that a protein
compound was chemically formed ...
—*Charles Darwin, letter to J. D. Hooker, 1 February 1871*

At the heart of the central issue of the origin of life and genetic
information lies Schrödinger's simple question, 'What is life?' or
more precisely, 'What is the physical nature of life?' In essence life is a
highly complex carbon-based chemical system that is maintained far
from chemical equilibrium by a constant influx of energy. The rates of
the chemical reactions maintaining the flux of metabolites in a cell are
enhanced, according to the Law of Mass Action, by the utilisation of
high local concentrations. These can take the form of absolute high
concentrations, the precise localisation of a reaction substrate at the
reaction centre of an enzyme and the catalysis of successive reactions in
a pathway by an array of catalytic centres, ordered on a physical
platform – effectively a module acting as a factory. But which of these
characteristics can reasonably be regarded as fossils reflecting the primitive origins of life chemistry and which simply conserve subsequent
crucial innovations in life's evolution? Evolution is of course a largely
continuous and incremental process but is punctuated by 'emergent'
events that may provoke a step change in evolutionary trajectory. An
excellent recent example of such an event would be the appearance of

two-winged insects – the Diptera or flies – in which a simple mutation transformed the hind wings into balancing organs – halteres – allowing the newcomers to exploit new modes of living. But is this concept relevant to life's origins? Theories of the origins of life are legion, and many depend on the more narrow, albeit essential, issue of the appearance of a suite of molecules necessary to maintain primitive metabolic processes. Nevertheless, two central, but related, ideas have dominated the debate. One is the notion that the essential components of life originated in a 'warm little pond', possibly maintained by geothermal energy and effluvia. The second again posits an essential role for geothermal energy but places the cradle of life within or in close proximity to undersea volcanic vents.

In these scenarios an obvious emergent event – in every sense – would be the escape and dissemination of primitive cells from the formative nursery. Such cells, as for present-day cells, would be self-contained entities enclosed by a selective barrier or membrane. Most importantly, they would likely possess all the principal characteristics that we associate with present-day life forms, albeit perhaps in a simpler form. The chemical systems would be complex, and thus, by definition, be of modular organisation (see Box 4.3). Further, this chemical complexity would be far from equilibrium and maintained by a readily available energy source. In this view complexity is a property of the pre-emergent state and thus should reflect both its prior evolution and the prevailing environmental regime in 'the cradle of life'.

But what drives the evolution of biological complexity? Is it simply a system in which Darwinian natural selection depends on the survival of the 'fittest' – or from the perspective of energy, the survival of the current most energetically efficient form appropriate to a particular environment? The evolution of green plants from single-celled algae to some present-day staple crops provides an instructive example. Today the most highly efficient converters of carbon dioxide into fixed carbon using light energy are C_4 flowering plants – so-called from the use of certain metabolic pathways. C_4 plants include maize, sorghum and sugar cane, although not rice, and have been estimated to utilise light energy at an approximately tenfold greater rate than unicellular photosynthetic organisms (Table 7.1).

C_4 metabolism is a relatively recent evolutionary innovation and has likely arisen on multiple occasions in C_3 angiosperm lineages, over

Table 7.1 *Evolution of photosynthetic efficiency*

	Age (Mya)	Energy flow (normalised)
Unicellular photosynthetic protists	>470	1
Gymnosperms	350	5
C_3 angiosperms	125	7
C_4 angiosperms	30	10

Adapted from Chaisson (2011), with permission from John Wiley & Sons, Ltd. Copyright © 2010 Wiley Periodicals, Inc.

which it has a competitive advantage under conditions of drought and high temperature. However, C_4 metabolism does not only result in a change in the favored metabolic pathway but also in a change in leaf structure – effectively altering the complexity of the system by recasting the underlying modular organisation. Increases in photosynthetic efficiency can, of course, also arise from other innovations such as improved mechanisms of photon harvesting. All these innovations represent an increase in available energy within the environmental constraints of the system.

There are, in addition to a genetic codescript, particular fundamental attributes that potentially provide important clues to the origins of life. Two stand out. First, the ionic milieux outside and inside contemporary cells differ substantially. On the exterior of a cell sodium is by far the most abundant monovalent cation relative to potassium, but inside their concentrations are essentially reversed (Macallum, 1926). Allied to this difference in the Na^+/K^+ ratio is an imbalance in the hydrogen ion concentration (pH) across certain membranes. Here the external milieu is more acidic than the internal – a difference that drives one of the fundamental mechanisms for the generation of ATP, the biological mediator of energy transduction (Lane, Allen, & Martin, 2010). Such pH differences are manifest across the membranes of both free-living prokaryotes and also in their role as symbiotic commensals (mitochondria and chloroplasts) in eukaryotic cells. The absolute hydrogen ion concentrations involved are several orders of magnitude lower than the Na^+ and K^+ concentrations (although the external/internal ratio is not). Notably both in many prokaryotes and within eukaryotic cells the medium immediately external to the energy-producing membrane is often spatially buffered by an additional structure.

The second universal, and likely primitive, property of living systems is the extreme crowding of macromolecules in the interior of a cell. In particular both in sperm and in virus particles where the preservation of the integrity of the DNA codescript is paramount, DNA condensates approach the limits of molecular compaction. Indeed, in the 1930s, Alexander Oparin proposed the idea that the first step in the origin of life would be the phase separation of these macromolecules into liquid coacervates (Oparin, 1936; Lazcano, 2010) (see Box 4.2).

Latterly the concept has been extended to suggest that in living cells the coalescence of different sets of molecules into distinguishable liquid compartments resulted in the establishment of a subcellular architecture in which different reaction centres – and their substrates – are spatially restricted (Lazcano, 2010). Although the chemistry of these compartments can be, and often is, highly dynamic, the modular structure is typical of complex systems and is not necessarily dependent on interactions of high chemical specificity.

Both these properties – ionic differences and macromolecular crowding – necessitate an interface or barrier between the outside and inside of a cell. Today this function is provided by membrane comprising a lipid bilayer containing both ion-specific and small-molecule-specific channels. The question is to what extent the origin and subsequent evolution of the twin central properties are a consequence of the same process.

Inevitably discussions of life's origins are highly speculative, but they should perhaps be judged by their accommodation to the characteristics of a complex system, especially including a mechanism for generating local heterogeneity in the concentrations of crucial reactants. Among the varied proposed scenarios are terrestrial 'warm little ponds', one of Darwin's more prescient speculations, and submarine hydrothermal vents, often located in the vicinity of the mid-ocean ridges (Box 7.1).

Both marine smokers, of both colours, and terrestrial 'mud-pots' can, in principle, provide porous structures that could act as an interface, or at least a partial barrier, between distinct aqueous milieux – the hydrothermal fluid and an external sea or more extensive pond. In white smokers such structures are calcareous, while in mud-pots they would

Box 7.1 Possible Geothermal Driven Origins of Life

Hydrothermal vents are essentially undersea geysers where hot fluid – often at a temperature in excess of 200°C – is vented through chimney-like structures on the sea-bed. These vents, or 'smokers', are most frequently located at mid-ocean ridges and are generally classified into two broad types – black smokers and white smokers, so-called from the colour of the emerging fluid. Black smokers occur around the centre of ridge, while white smokers are found on the flanks. Black and white smokers differ in several crucial respects. The fluid in black smokers is rich in iron and sulphur, leading to the deposition of black iron monosulphide and is also acidic. In contrast the lower temperature white smoker fluids are more alkaline and deposit calcium-based chimneys. In both cases the sodium/potassium balance is similar to that of sea water, i.e., sodium rich.

Terrestrial 'warm little ponds', aka geothermal fields, often overlie magma sources close to the surface. The Yellowstone area in the United States is a well-known example. Extant fields differ widely in chemical

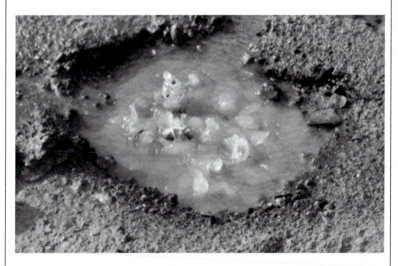

Figure 7.1 Darwin's warm little pond? A Kamchatka mud-pot. *A black-and-white version of this figure will appear in some formats. For the colour version, refer to the plate section.*
Source: Reproduced from Dibrova, Galperin, Koonin, & Mulkidjanian (2015) with kind permission of Anna Karyagina

character and geological provenance. Some, such as Yellowstone are sodium rich, while others, exemplified by a fumarole 'mud-pot' swarm in the Kamchatka Peninsula, are potassium rich (Figure 7.1). In contrast to conventional hot springs, mud-pots result from the release and subsequent condensation of superheated steam enriched in siliceous minerals and a variety of metallic cations. The precise ionic composition would be dependent on the local geology. For example, some rocks are potassium rich and others potassium poor. Notably the early continental crust that was prevalent at least 4,200 Mya ago in the era was potassium rich and could have been a source of potassium-rich vapours (Boehnke et al., 2018).

be likely siliceous. Not only would the mineral skeleton provide a surface for concentrating reactants, but also it could participate catalytically in the synthesis of essential precursors for the synthesis of biological macromolecules. But, most importantly, by acting as an interface with the world outside, the evolving complex prebiotic system could adapt to external changes. For both terrestrial mud-pots and marine smokers, energy and essential reactants would depend on geothermal activity. But to achieve today's ionic imbalance in the K^+/Na^+ ratio between the intracellular and extracellular solutions would require adaptation from either a high K^+ state or a high Na^+ state. While marine smokers are, by definition, surrounded by the Na^+ rich sea, in terrestrial geothermal fields the K^+/Na^+ ratio of the liquid phases depends on their origin. The fluid that condenses from the superheated vapours in the mud-pots of a geothermal field can be potassium rich, while, in contrast, the ascending cooler marine hydrothermal fluid is likely Na^+ rich. The condensate is also likely to be more alkaline (a lower hydrogen ion concentration) than the hydrothermal fluids. Both alkalinity and higher potassium concentrations would be consistent with the geologic conditions prevailing in the Hadean era and coupled with a contemporary or subsequent serpentinisation (Boehnke et al., 2018). Again, certain siliceous formations with hollow tubular architectures could, in principle, provide a scaffold for directional ion transport (Glaab, Kellermeier, Kunz, Morallon, & Garciá-Ruiz, 2012). With the establishment of a barrier, any ionic changes in the external fluid – by, for example, the infiltration of more Na^+ rich waters, and absorption of carbon dioxide from the atmosphere – could be better accommodated without disruption of any preexisting K^+ dependent complexity. Definitely a warm little pond, but one perhaps slightly different to that conceived by Darwin.

The concept of a barrier is also highly relevant to one of the more fundamental properties of today's living cells – the generation of energy in the form of ATP by the flow of hydrogen ions across a membrane. There is no reason to assume that the absolute salt concentrations outside and inside a barrier would be identical – indeed, it would seem unlikely that they were so. If they were not there would be a difference in electric potential across the active interface, which would drive a flow of ions through the barrier. For example, if the exterior were more acidic, hydrogen ions could pass inwards together with sodium ions, while potassium ions would pass outwards. Such a scenario would be consistent with the separation of an interior compartment containing the vapour condensate in a mud-pot from an exterior domain derived from the hydrothermal fluids.

The initial evolution of biological complexity in a potassium-rich environment is clearly only one of perhaps several possibilities. An alternative scenario would be a nursery in the sodium-rich environment of an undersea smoker, followed by a gradual transition to the present-day internal (intracellular) potassium-rich milieu. In the context of creating a complex system this would be the 'Tempus' solution (see Box 2.1). An initial chemical evolution in a sodium rich environment would require many linked adaptations to convert to a potassium dominated chemistry and so would have a much lower probability in an essentially Bayesian world (see Danchin & Nikel, 2019 for a discussion).

Barriers also provide a mechanism for achieving high local concentrations (Figure 7.2). The architecture of vents could allow concentration at dynamically maintained bounding structural constraints. This contrasts with some more static pictures of 'warm little ponds' that envisage concentration by evaporation followed by recharging (Pearce, Pudritz, Semenov, & Henning, 2017).

There are two other features that follow from the existence of a 'semipermeable' barrier. Most importantly conditions at the interface between the vent fluid and the external medium would be far from

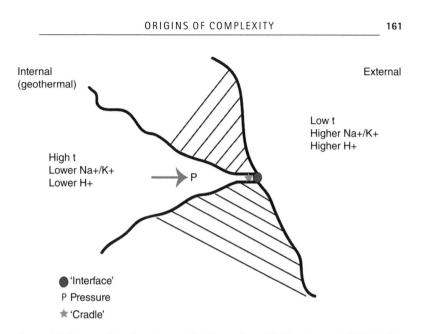

Figure 7.2 Cartoon of interface between K$^+$-rich geothermal fluid and external Na$^+$ fluid. The cartoon shows how a semi-permeable barrier between a K+ rich geothermal fluid and an Na+ rich external fluid (e.g., a sea or a local runoff pond) could establish conditions necessary for energy generation and local organisation.

chemical equilibrium and would be maintained by a continuous influx of vent fluid. This physicochemical condition at the interface thus coincides with a primary physical requisite of life. The second, less obvious, feature is that the internal and external temperatures would differ (the external fluid would be cooler), possibly substantially, resulting in a highly localised temperature gradient in the vicinity of the interface. This could facilitate phase transitions such as the formation of coacervates and consequent concentration of interacting molecules. Above a certain size any such concentration could then become trapped at the barrier and the high concentrations maintained by pressure exerted by the vent fluid.

In summary, many of primary characteristics of life are intrinsic to the concept of an active interface between hydrothermal vent fluids and a compositionally distinct external fluid – be it marine or terrestrial. Applying Ockham's razor, this postulate is sufficient. Clearly this model implies that strict physical limits be placed on the dimensions and nature of the putative interface.

Another essential and related consideration is the nature of primitive metabolic pathways (Adam, Borrel, & Gribaldo, 2018). Miller (1953) and subsequently Miller and Urey (1959) first demonstrated that first simple and then more complex organic molecules could form in environments that are thought to mimic early earth and could be derived from simple physical processes. These observations have been amply confirmed and extended by more recent experiments, although the precise nature of the original precursor molecules, for example, nucleotides, remains a matter of debate. Both undersea hydrothermal vents and small little ponds can, in principle, provide appropriate conditions for the genesis of biologically important precursor molecules and also for the subsequent synthesis of related macromolecules.

Genetic analysis suggests that an ancient process for the conversion of simple organic molecules into energy is the Wood–Ljungdahl pathway (Figure 7.3) (Williams et al., 2017; Adam, Borrel, & Gribaldo, 2019), which, in extant organisms, catalyses the conversion of carbon dioxide to essential intermediary metabolites under reductive – essentially oxygen-free – conditions, possibly equivalent to those prevailing in a particular type of hydrothermal vent. Today, many of the chemical reactions in this pathway require the participation of a small molecule, acting as a coenzyme bound to a protein – together constituting the enzyme. One role of the protein moiety of the enzyme is to bring the coenzyme and the substrate metabolite into close spatial proximity so increasing the rate of the resultant chemical reaction. But because the generation of the protein components is by itself a complex process, simplicity suggests that the panoply of coenzymes associated with the Wood–Ljungdahl pathway – and others – are likely primitive, probably even more primitive than their current protein partners. Although the reactions they catalyse are remarkably varied, the chemical nature of the coenzymes is revealing. Most (Box 7.2), with the exception of coenzyme M, manifest a ring structure or structures – often heterocyclic. This, of course, is a characteristic they share with nucleic acid precursors. Electron delocalisation in heterocyclic rings can enhance reactivity, but the array of similar structures might also co-congregate and thus co-localise on appropriate inorganic or carbon-based organic platform. There is no evidence that such platforms existed in prebiotic nurseries, but a simple current example would be the high propensity of compounds of this type to be co-adsorbed by activated charcoal. Although conjectural, by bringing

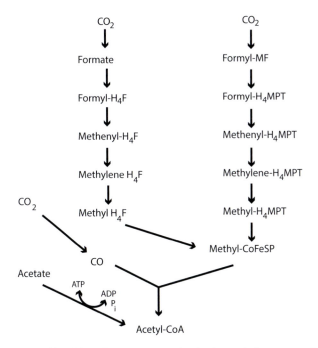

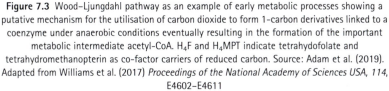

Figure 7.3 Wood–Ljungdahl pathway as an example of early metabolic processes showing a putative mechanism for the utilisation of carbon dioxide to form 1-carbon derivatives linked to a coenzyme under anaerobic conditions eventually resulting in the formation of the important metabolic intermediate acetyl-CoA. H_4F and H_4MPT indicate tetrahydrofolate and tetrahydromethanopterin as co-factor carriers of reduced carbon. Source: Adam et al. (2019). Adapted from Williams et al. (2017) *Proceedings of the National Academy of Sciences USA, 114,* E4602–E4611

together a mixture of metabolic catalysts platforms of this nature could be regarded as primitive precursors to biological complexity. The association of heterogeneous catalysts in close spatial proximity could impart directionality on a chemical process by providing successive substrates in a reaction pathway. Such a system would not need to be in chemical equilibrium and represents an energy-efficient mechanism for the production of more complicated molecules.

The Origin of Nucleic Acids

The topic of the origin of nucleic acids is fundamental to considerations of biological information. The nature of the chemical precursors of the nucleic acids bases is somewhat controversial – hydrogen cyanide or

Box 7.2 A Gallery of Coenzymes

Coenzymes, in some cases aka cofactors, are organic chemically reactive molecules that can act as intermediates in enzymatically catalyzed reactions. Such reactions include many oxidation-reduction couples and one- or two-carbon atom transfer. Present-day coenzymes are often chemically complicated – for example, Cofactor F430 and cobalamin (vitamin B_{12}) but Coenzyme M is an exception. Whether this complexity is primitive or subsequently adaptive is unclear. Many of the steps in the Wood–Ljungdahl pathway (Figure 7.3) currently utilise coenzymes, notably THFA, THMPT, Coenzyme A, Coenzyme B, Coenzyme M and Cofactor F430. Of these, all except Coenzyme A are involved in one-carbon atom transfer. Other coenzymes, including flavin adenine dinucleotide (FAD), nicotinamide adenine dinucleotide (NAD) and Coenzyme Q, are currently components of the electron transport chain coupling ATP production to a hydrogen ion gradient. With the exceptions of Coenzymes B and M, all contain aromatic heterocycles and negatively charged groups including phosphate, sulphite and carboxylate,

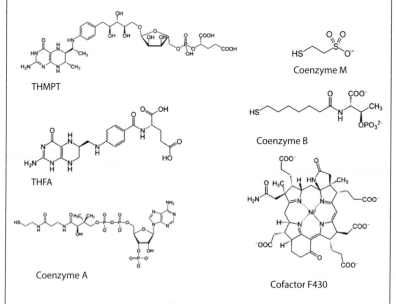

Figure 7.4 Representative coenzymes (cofactors) utilised in the Wood–Ljungdahl pathway. THFA, tetrahydrofolate; THMPT, tetrahydromethanopterin.

> just like nucleotides. Indeed, some (e.g., Coenzyme A, FAD and NAD) contain a nucleotide moiety. The problem of the origin of coenzymes is thus analogous to that of nucleotides. In Coenzymes B and M the one-carbon residue is bonded to the sulphur atom as also in S-adenosyl methionine, another one-carbon donor.

formamide? (Saladino, Botta, Bizzarri, Di Mauro, & Garciá-Ruiz, 2016; Saladino, Di Mauro, & Garciá-Ruiz, 2019; Sutherland, 2016) – as also is the manner of their assembly into a polymer. Were the bases polymerised when free in solution or on a solid template? This distinction is essentially analogous to the primitive evolution of metabolic pathways catalyzed by coenzymes. By providing a stable platform a solid template would enable the placement of precursor bases in close spatial proximity, potentially lowering the activation energy for the formation of internucleotide bonds, and so acting as a catalyst (Saladino et al., 2019). Notably certain silica-based self-assembled structures provide a mechanism for compartmentalisation and would be fully compatible with a silica and K^+ rich scenario for the origin of life.

Once formed, a small RNA chain could, in principle, by base-pairing, act as a template for the synthesis of its complement. And, in turn, the complement could be a template for the synthesis of the originating sequence. Such a process, at least initially, is unlikely to be precise and could result in the production of an immense variety of different sequences, which themselves would likely be subject to Darwinian selection. One such consideration would be stability. Because the individual bases are chemically distinct, any resultant polymeric chains would in principle differ in chemical stability depending on their base composition. Indeed, in the test tube complex, sequences resist degradation by hydrolysis markedly more than monotonous ones, thus potentially favoring the evolution of sequence-based genetic information (Ciciriello et al., 2008). Similarly in contrast to monotonous sequences in complex sequences would potentially be able to adopt a variety of folded structures mediated by, possibly, initially very transient base-pairing paving the way for more selective intermolecular recognition.

The acquisition of folded structures or the potential to form them would be prerequisite for the development of intramolecular catalytic activity – aka a ribozyme (Cech, Zaug, & Grabowski, 1981;

Guerrier-Takada, Gardiner, Marsh, Pace, & Altman, 1983) – by an RNA molecule. Such catalytic activities could be used in intramolecular or intermolecular cleavage events to yield more functional RNA molecules. This property has considerable implications for scenarios for the early evolution of informational macromolecules. If RNA can act as a catalyst then, by reducing any requirement for protein catalysis, the origin of the genetic code would play a much less critical role in the early stages of the first biochemical systems that were capable of replicating themselves (Crick, 1968; Orgel, 1968; Woese, 1968). Ultimately ribozyme activity would be directly engaged in protein synthesis to catalyse the formation of a peptide bond. Again, it is an evolutionary progression to a more organised system.

Escape from the Nursery: RNA Alone or DNA Plus RNA?

Modularity, as manifested by intracellular compartmentalisation, is a fundamental attribute of extant complex biological systems. An extreme example is the functional and spatial separation of RNA and DNA in all living cells. Although there is some overlap, in general DNA, either in the prokaryotic nucleoid or the eukaryotic nucleus, is highly compacted into a small volume, while in contrast, RNA molecules including principally tRNA, mRNA and rRNA involved in the active process of protein synthesis occupy a separate and distinguishable intracellular domain. As has been understood for well over half a century, the distance between the centres of individual duplexes in condensed DNA (normally in the range of 20–50 Å) is insufficient to accommodate some molecular machines, such as RNA polymerase, central to genetic information flow. For many RNA molecules such condensation would be anathematic to their function. However, the relative potential compactability of DNA and RNA depends in part on both their intrinsic chemical differences and on the nature of their interacting proteins. For example, polypeptides enriched in alanine and lysine can drive DNA condensation by charge neutralisation, and in starved bacteria an abundant DNA binding protein selectively interacts with DNA toroids. Additionally, the compaction of DNA, either as the free molecule or in combination with basic proteins, is strongly dependent on its ionic environment. Notably compaction is promoted by K^+ to a much greater extent than Na^+ (Wu & Travers, 2019). The physical basis for this phenomenon is simple. At comparable concentrations K^+ increases DNA twist substantially more than Na^+ (Anderson & Bauer, 1978). Certain divalent cations, for example Mn^{++}, prominent in hydrothermal environments, have a similar effect. This is possibly another argument for the primeval ionic composition influencing the evolution of pre-biotic chemical complexity.

Condensation of DNA-protein complexes not only reduces the volume occupied by a suite of molecules but also facilitates compartmentalisation of structure and function. Thus in contemporary nuclei, phase separation in a liquid system is responsible for the generation of distinguishable compartments – examples include nucleoli for the dedicated synthesis of ribosomal RNA, transcriptionally competent euchromatin and transcriptionally repressed heterochromatin. These represent a very different level of organisational complexity from a simple homogeneous solution mixture. In present-day cells liquid–liquid phase separation is often associated with simple peptide sequences composed of a limited number of amino acids. These peptide sequences, termed intrinsically disordered peptides (IDPs), are structurally very dynamic, i.e., they are characterised by a relatively high intrinsic entropy. By their very nature their chemical interactions with a ligand are repetitive and lack precise selectivity – for example, the interactions of certain basic polypeptides found in chromatin proteins with DNA. As argued in Chapter 4, such sequences could represent an early stage in the expansion of the genetic code and are likely primitive (Figure 7.4). Only subsequently with the accession of more amino acids to the genetic code would intramolecular recognition between biological macromolecules become more precise. In this view compartmentalisation would represent an early stage in the evolution of biological complexity and be a necessary and essential precursor to the more ordered complex systems existing today. Put another way, spatial heterogeneity, implying functional distinction, is a dominant characteristic of biological complexity. A further consequence of early compartmentalisation would be the establishment of delimited volumes where reactants could be selected and concentrated, thus increasing the efficient utilisation of energy. Energy and information would be required to select and concentrate reactants, but in principle this is no different – just more complicated – to the selection of gas molecules on the basis of their energy state by Maxwell's demon.

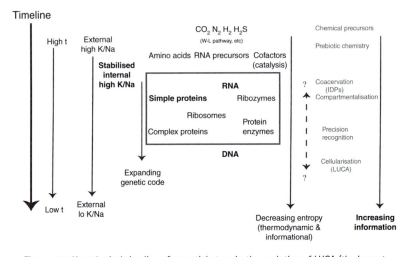

Figure 7.5 Hypothetical timeline of essential steps in the evolution of LUCA (the Lowest Universal Common Ancestor). Escape to autonomous living would follow or be contemporary with cellularisation. The 'RNA world' is outlined. IDPs indicate intrinsically disordered and probably repetitive peptides (for discussion, see also Chapter 4).

RNA and DNA are not the only biological macromolecules able to form coacervates. Certain types of short polypeptide chain, known as amyloids, form condensates with a fibrillar morphology reflecting an underlying one-dimensional crystal structure. Although notorious for their role in neuro-degenerative disease, these condensates can possess enzymatic activity and can act as templates for their own replication from amino acid precursors. As a consequence, amyloids have been suggested as alternative candidates to nucleic acids for the transmission of 'genetic' information.

A fundamental question – to which there is currently no answer – is whether the appearance of DNA as a genetic repository preceded the escape of autonomous primitive cells from a founding nursery. If so, a world with genetics but no DNA – the 'RNA world' – would have been a relatively transient stepping stone to a higher biological complexity (Figure 7.5). For an autonomous free-living cell, a separate condensed genomic compartment could confer significant selective advantages – notably portability and the availability of expansion by, for example, genomic duplication. Nevertheless, although the spatial separation of DNA and RNA is an excellent example of compartmentalisation and arguably sufficient for escape, the evolution of complexity by intracellular compartmentalisation is likely an ongoing process. Even within the eukaryotic nucleus the site of rRNA synthesis, the nucleolus constitutes a separate spatial domain.

How can this scenario – as well as other related ones – account for the genesis of biological information? Put another way, did the appearance of biological 'information' precede the evolution of a codescript, and if so, how can the transition be described? At its simplest 'information' is simply a measure of the organisation or complexity of a physical system. The speculative scenario outlined earlier implies that the genesis of the chemical system that we call 'life' lay in the fortuitous assembly of concerted chemical reactions – or metabolism – catalysed initially by small molecules – coenzymes – and only subsequently by more complex templates – enzymes. Such a physical system would possess 'information' but would not, by itself, require a codescript. Nevertheless, the development of a codescript in a prebiotic environment could initially act to stabilise the chemical factory via, for example, short peptides and then evolve to encode some or all of the 'information' contained in the system. This emergent step would be the basis for life as we know it, providing a mechanism for the conservation and replication of biological information in a compact set of instructions analogous in some ways to those for building a large Lego model. This step would also be a prime example of an increase in complexity – system stabilisation accompanied by an increase in chemical efficiency, which could then serve as a substrate for further evolution.

To illustrate how coding could evolve, a simple analogy would be to consider the context of describing the classic box presentation of the genetic 'code' (see Figure 4.3). This table can be interpreted from two different perspectives – as a description of isolated and independent chemical interactions between a codon and its anticodon contained in a tRNA or as the consequence of the translation of a linear array of codons in an mRNA into a defined polypeptide. It is this latter context that is normally implicit in most discussions and is generally, and possibly universally, assumed to be the basis for 'genetic information'. In isolation, base-pairing between a codon and its anticodon is equivalent to an 'informational interaction' of the type described in Chapter 2. Although base-pairing is highly specific, it is really chemically no different in principle from the many other specific interactions that are assumed to have taken place and defined the complexity of the metabolism of a

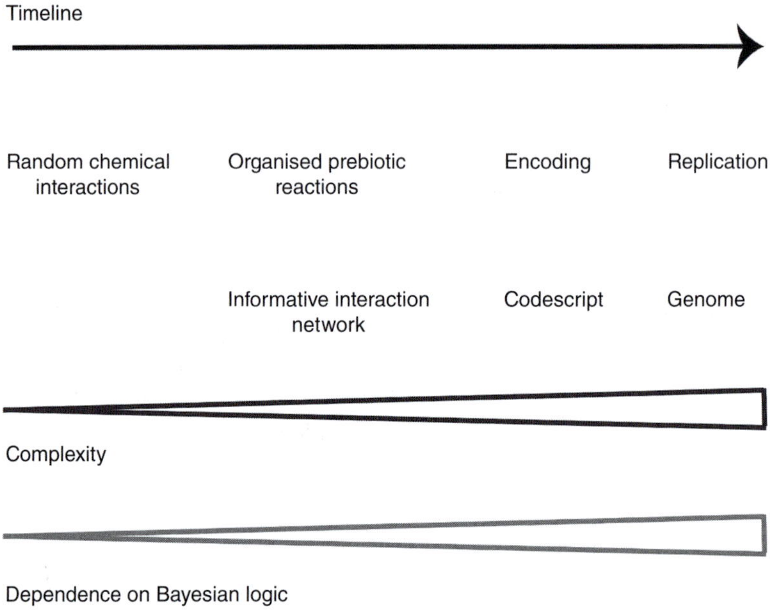

Timeline

Random chemical   Organised prebiotic   Encoding   Replication
  interactions   reactions

                           Informative interaction   Codescript   Genome
                                 network

Complexity

Dependence on Bayesian logic

Figure 7.6 Genesis of biological 'information'. See text for details.

prebiotic nursery. Such isolated interactions do not by themselves constitute a 'code'. However, with the probable and possibly parallel advent of
tandem arrays of codons – even as short as two – 'codescripts' were born
requiring successive codon–anticodon interactions for the generation of a
defined and simple polypeptide. This emergent transition would thus
pave the way for the evolution of copying processes including transcription and replication and of course genetic inheritance. In this view the
ultimate encoding of the early informational interactions into a conserved
codescript is a direct consequence of the further chemical evolution of an
already complex system (Figures 7.5 and 7.6). This process would favour
the most energetically efficient outcome.

8

The Complexity of Societies

The final conclusion ... is that casual groups are never an object of [natural] selection, but, social groups, as cohesive groups, may indeed be a target of selection.
—Ernst Mayr, 1963.

A central issue in modern biology is the question of the mechanism of evolution of social groups. Such groups are considered to be integral components of biological complexity (Mayr, 1963), and the discussion of this point has been one of the most animated – and often the most sterile – debates in biological thinking. It is often framed as a distinction between kin selection, driven by altruistic behaviour by members of a family group sharing a panoply of genes, and group selection where the cohesion of the group, and not necessarily genetic relatedness, is the driving factor.

But animal groups are not all the same. In casual groups individuals of the same species tend to congregate together at certain times, whereas social groups are more generally cohesive and have a greater organisation. An essential difference between a casual group and a social group is not size, but complexity (Mayr, 1963). A murmuration of starlings – those graceful swirling clouds gyrating in the evening sky – is a homogeneous assembly of individuals with no obvious differentiation. In contrast, within social groups of all types, different individuals perform different tasks. Within insect communities, reproduction, foraging and

---

Box 8.1 Starling Murmurations

The simple behavioural pattern of a starling murmuration results possibly from the coordination of visual patterns among its participants. Its purpose, if any, is obscure. Possibilities would include keeping warm en route to the evening roost, confusing predators and communicating the findings of the day in a fashion analogous to social corvids.

---

defence are performed by distinct, and often morphologically distinguishable, castes. Although insect colonies may comprise tens, or even hundreds of thousands, of individuals, the variety of functional distinctions is quite small. Human societies present an entirely different picture. Narrow skill specialisation is usually the norm, resulting in a large array of essential skill-sets being brought together within a community. In terms of cooperation, a human society is more complex than an insect society, which in turn is more complex than a murmuration of starlings (Box 8.1).

Is it legitimate to use the term *complexity* in the context of animal groups in the same way as in the contexts of biological molecular systems or even ecosystems? If so, the implication is that more complex social groups have evolved by the acquisition of more information and that that information is selectable. Writing on this topic, Mayr (1963) succinctly stated, 'The final conclusion ... is that casual groups are never an object of [natural] selection, but, social groups, as cohesive groups, may indeed be a target of selection. To qualify as a potential object of selection, a social group must be clearly delimited and compete with other social groups'. Ultimately selection acts on a phenotype that is specified by information arising from DNA, but not necessarily excluding an additional source. For group selection this information must be contained within the group and must act to promote the cohesion of a diversity of individuals.

It is pertinent here to ask whether there are parallels between biological complexity and the complexity of animal, and particularly human, societies. A major study on the rise and fall of human civilisations by Joseph Tainter (1988) described their complexity in terms of five major characteristics. First, as with all complex systems, societies, and their associated sociopolitical systems, require energy for their maintenance. Similarly, like biological systems, more complex societies are more heterogeneous. These requirements are correlated with increasing costs per capita as complexity increases such that investment in increased complexity as a sociopolitical response often reaches a point of declining marginal returns. Importantly, increased complexity is associated with parallel increases, often massive, in the information content of the social system as a whole. Indeed, our lack of knowledge of periods following the collapse of highly organised societies – the so-called dark ages – can at least be partially ascribed to a dearth of accessible informational records.

For human societies it is common observation that there is a trend for organisational units to increase both in population and in the addition of administrative hierarchies. These trends, paralleling advances in biological organisation, are prominently related to energy sources. Thus the transition from a hunter–gather mode to a more settled agricultural existence provided a necessary condition for the evolution of more complex civilisations (Renfrew, 1987). It is likely no accident that the earliest such civilisations of which we have detailed knowledge arose on fertile alluvial soils. The establishment of the Mesopotamian, Harrapan and Egyptian civilisations on the rich soils provided by the waters of, respectively, the Tigris/Euphrates, Indus and Nile rivers, is suggestive in this context. Similarly, the very early Xia dynasty in China arose on the highly fertile loess soils bordering the Hwang-ho River. Subsequently, the availability of alternative easily available sources of fossil fuel – coal in the first instance and then oil and gas – could have fuelled an explosive growth in societal complexity culminating in the late development of supranational entities. To put this into an evolutionary context, the rise in energy flux associated with the development of complex societies is quantitatively comparable to that with the preceding elaboration of biological complexity (Chaisson, 2013) (Figure 8.1), yet accomplished over a 10^4–10^5 times shorter timescale. Just as biological systems evolve more efficient, and sometimes emergent, energy harvesting mechanisms so also can civilisations. While novel energy sources provide a permissive environment for this rapid acceleration, it is the enabling of their utilisation that is crucial. Essentially this implies an increase in available information is accompanied by a substantial increase in entropic export to the total physical system. The principle driving the rise of civilisations

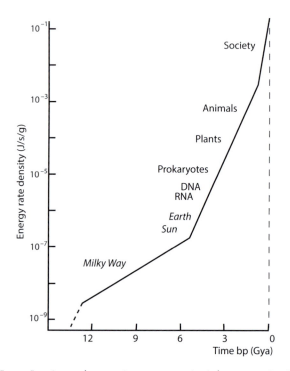

Figure 8.1 Energy flux changes (expressed as energy rate density) accompanying the evolution of complexity. As the complexity increases, so also does the energy rate density. Biological and cosmological complexity levels are shown, respectively, in Roman and italic type. Energy flux – expressed here as J/s/g – is a measure of the rate of energy consumption by the system under consideration. Source: Based on Chaisson (2011), with permission from John Wiley & Sons, Ltd. Copyright 2010 Wiley Periodicals, Inc.

is exactly that initially propounded by Boltzmann for, as he saw it, Darwinian evolution.

Schrödinger argued that the genetic codescript – and hence its information content – was essential for the entropic minimisation inherent in biological organisation. Does the same principle apply to societal complexity? Arcane discussions of the evolution of societal organisation are often predicated on a distinction between 'kin' – essentially genetic – and the less rigorously defined 'group' mode. One basis for this distinction is the altruistic behaviour of non-reproducing individuals within a colony in which all the individuals are closely related. A honeybee colony is a classic example of such a eusocial organisation, which is also found in termites and mole-rats. Here the non-reproducing female workers nurture the brood of a single queen. (As to whether honeybees are indeed a paradigm of kin selection is controversial [Box 8.2]). Given that the examples of altruism probably represent only a fairly minor component of the whole spectrum of social behaviour, in the context of information the broad distinction between genetic information and the

---

Box 8.2 Altruistic Behaviour

Altruistic behaviour is deemed to aid the survival of the DNA of closest relatives (see Dawkins [1976] for an extensive background argument), and therefore the propensity for altruism depends on the degree of relatedness of individuals. A quantitative theory of altruistic behaviour is based on this criterion (Hamilton, 1964). However, in the example of honeybees, a queen can mate with up to at least 30 drones. Consequently, as pointed out by E. O. Wilson (Nowak, Tarnita, & Wilson, 2010, and references therein), the degree of relatedness between sister workers falls below the threshold deemed to define 'kin' selection, and therefore bee behaviour should be more correctly classified under the umbra of 'group' selection. In the theory the degree of relatedness among sister workers rests on the assumption of equal weights being assigned to the queen and drone components of the inherited DNA. But is it just the theory of kin selection rather than the overarching principle that is flawed? If the input values for relatedness were skewed to give, for example, more weight to the queen in bees, the apparent contradiction disappears and can be extended to other eusocial animal societies in which the unusual hymenopterid condition of haplodiploidy (diploid queen and workers but haploid drones) does not occur. Such an extension, of course, would possibly include social groups with no altruistic characteristics. Ultimately, it is likely the chemistry of volatile organic compounds such as pheromones that determine altruistic behaviour in bees, as also social behaviour in African hunting dogs and mole rats. The caste differentiation and behaviour exhibited by honeybee colonies are determined primarily by the interaction of epigenetic information and pheromones. Both of these are largely genetically determined and represent the outcome of selection on genes.

ultimately cultural information seems a simpler and more sustainable option. While there is a demonstrable correlation between societal complexity and information availability, a causal connection is difficult to establish. Nevertheless, the early Egyptian, Mesopotamian and Achaemenid civilisations were all noted for their accumulation of written records, at least partly for revenue-gathering purposes, and thus for the maintenance of a complex organisation. Indeed, so important was such information that it was encoded on the heads of Darius' messengers (Holland, 2005) – only to be recovered on their return to Persepolis. Initially writing and its interpretation would have been a largely esoteric occupation but with the invention of the printing press by Gutenberg and its further popularisation by Caxton, the printed word became more accessible. In England William Tyndale's advocacy of a Bible for everyone in the sixteenth century further hastened this trend. Latterly the advance of literacy, coupled with the relative permanence of the printed book and finally, the explosion of electronic media have combined to increase the amount and reach of available 'information' in a society. Associated with this explosion is the rise of information-processing organisations, including supranational polities (of which the European Union is a paradigm) and resource-managing institutions, which often operate beyond the effective reach of national and supra-national control. These activities are manifestations of additional hierarchies of complexity and, as such, can be resource consuming. In some cases maximisation of profit comes at the expense of resources or real wealth. A parallel in biology would be ash dieback disease.

Written information by itself is not genetically determined and so is entirely different from the DNA codescript. So what is its nature? In many, very diverse, animal societies children and adults learn from their parents and peers by example. This is likely a fundamental factor directing their evolutionary pathway. For example, corvids learn tool-using from others while the feeding preferences of individual orca pods are retained within a group. This is little or no different from human children being taught by their parents. For those that learn, the information gained is not genetic but is acquired – and can used and be passed on during their lifetime. This again, as in the example of angiosperm shoots, is the essence of Lamarckism and is thus completely distinguishable from the zoological paradigm of genetic inheritance, where inherited mutations are usually generated within and passed on through the germline. Put another way, cultural information is Lamarckian (Chaisson, 2013). Not only can this information function within a lifetime, but it can also be immortalised in words and so can, where appropriate, add to the total available information in a society. On this argument the evolution of complex societies is a Lamarckian phenomenon (Box 6.1).

An important aspect of cultural information is that it can be used as a vehicle for maintaining social cohesion. Human groups now very often exceed populations that are dependent solely on kin-groups, suggesting another mechanism favouring stability comes into play. For cohesion factual accuracy of cultural information is not initially a necessary requirement – after all that is not its purpose. Its purpose is to bind together a large group of individuals more firmly. This is often facilitated by the observance of ritual and recourse to the supernatural. But if the use of cultural information in this context accomplishes for a time a viable increase in complexity, there is likely an ultimate cost. Further increases in available information may mean that Max Perutz's dictum that 'In science, the truth always wins' could eventually expose any contradictions in the information underpinning cohesion or at the very least create intellectual and political conflicts.

Do biological systems also exhibit the phenomenon of declining marginal returns? Here the argument is complicated by the hysteresis of responses to different modes of information transmission. For genetic responses timescales of a single generation are necessary before natural selection can, and probably does, act. For cultural information this may, although not necessarily, extend to several generations, if only because human biology is such that cultural information is accumulated over lifetimes much longer than the average generation. Additionally written and/or electronic information may have, especially for religious texts, lifetimes approximating to centuries rather than tens of years. Some cultural information – even if, as often, factually incorrect – may thus be more resistant initially to natural selection than genetic information while its other most recent manifestations may have more immediate outcomes. When challenged the maintenance of the supporting information may necessitate a greater resource input – otherwise a declining marginal return. Nevertheless, the phenomenon of declining marginal returns may also be apparent in purely genetic systems. A simple example is the *Escherichia coli* pangenome (Box 8.3).

Box 8.3 Pangenome

A pangenome, usually applied in the context of bacteria, especially the gut bug E. coli, is the total number of genes that can readily be exchanged amongst a population of the same organism – and in some instances closely related organisms. For example, while the *E. coli* chromosome characteristically contains about 5,000 genes, the number of transmissible available genes in the whole population may comprise as many as 15,000.

While in principle it might be possible to contain all ~15,000 genes in the pangenome required for a response to most foreseeable circumstances into a single circular chromosome, in practice the disadvantage of increased replication time is too great to offset the preferred solution of distribution of the same set of genes among multiple genomes. In terms of the number of genes in a chromosome, there will be a crux where any gains resulting from an increased gene (and therefore information) content are offset by greater costs. In other words, genes can be expensive. Similarly, although in principle more advanced vertebrate immune systems could be programmed to oppose a greater range of molecular threats, in practice the increase in complexity required would be expensive, and the system as a whole would invest more energy in this task than would be compatible with maintaining the efficiency of other necessary processes. It would be a case of complexity overreach.

A further analogy between societies and biology is the occasional loss of complexity. While, as argued earlier, provided the environment is benign, there is a tendency for complexity to increase by a positive feedback process, regression can occur. If the resources available, for whatever reason, become insufficient, the complex system becomes unstable and either collapses completely or reverts to a lower level of complexity accompanied by a functional loss. Ecosystems can exhibit such a transition. For example, when the indigenous trees in a rainforest are replaced by palm oil tress, the forest loses its capacity to regulate the local water cycle (Aragão, 2012; Spracken, Arnold, & Taylor, 2012). More extreme would be the periodic mass extinction events that have occurred throughout geological time. These events are manifested as a greater or lesser loss of biological diversity, which inevitably limits the number of available inter-specific interactions and consequently reduces complexity. Another possible, and debated, example is the genesis of cancers (Greaves, 2000). Here somatic mutations – often several – sever the regulatory link between a cell and its location within a multicellular organism. Having lost external regulation, the growth of a cancerous cell becomes, to a greater or lesser extent, independent of what is now its host. This transition involves a change, and arguably a loss, of some information that contributes to the complexity of the organism as a whole. It certainly results in a loss in interconnectivity and, in this respect, can be regarded as regression to a lower level of complexity. Nevertheless, a tumour is still a complex entity and may indeed, as originally suggested, be more complex at the molecular level than its progenitors. Yet it retains self-referentiality – in this sense it may represent only a regression and not a collapse. A cancer is thus one example of many where the costs of different hierarchical levels of complexity are rearranged. In societies collapse or regression of the most complex manifestations – for example, the Sardinian Nuragic civilisation (Melis, 2003) – does not necessarily imply obliteration – although under extreme circumstances it can. After such an event the society may still exist but functions at a lower, and usually less extensive, level of organisation. Such a regression would inevitably be accompanied by a lower energy input and information requirement. The loss of apparent complexity in Nuragic organisation can be attributed, at least in part, to their failure to transition from a thriving Bronze Age culture to the dominant Iron Age culture of their competitors. Possibly they lacked the information and/or the resources to make this emergent step.

In many ways the concept of the regression of civilised societies simply reflects the vulnerability of highly integrated complex systems. In simple obligate symbiotic systems exemplified by the fig/fig wasp and Darwin's moth/orchid (*Xanthopan/Angraecum*) associations, extinction of one specialised partner likely condemns the other. The robustness of these systems thus depends on the vulnerabilities of both components. In our modern world the principal driving force for the recent rapid expansion of 'globalisation' is probably economic efficiency. The resultant complexity depends on multiple highly interconnected supply chains analogous to a form of obligate symbioses between different societies. These supply chains represent a major vulnerability in the maintenance of globalised complexity.

Interruption, from whatever cause, can have severe repercussions. The recent COVID-19 pandemic, although perhaps a relatively minor perturbation in the long perspective of the more profound influences of cataclysms on biological complexity in the geological time frame, illustrates this well. The abrupt loss of interconnections – both physical and human – has both disrupted the world economic system and resulted in apparent loss of integration of cooperation between societies. Although it is much too early to anticipate the ultimate consequences of the COVID pandemic, the failure of integration could presage fragmentation. Put another way, the hierarchical complexity (see Figure 2.7) represented by transnational interactions has

Figure 8.2 Lamarck presiding over Le Jardin des Plantes, Paris. Photograph by Leon Fagel in the public domain.

the potential to revert to a more diverse lower level occupied by many disconnected nation states.

This discussion of the rise of societal organisation presupposes that societal complexity represents a seamless transition from molecular complexity specified by the DNA codescript. The independent rise of cultural information in other animals strongly supports this contention, and the process is thus likely dependent on genetic selection driving increasingly sophisticated information processing in the central nervous system. In this view societal evolution essentially is a consequence of the rapid evolution of DNA-independent information processing. There are arguments as to whether increases in societal complexity represent a continuum or exemplify punctuated evolution (Tainter, 1988). Both are characteristic of the natural world, and so either would be consistent with the paradigm. The crucial component in the rise of societal complexity is the transition from genetic to cultural information being the prime driver of evolution. The evolution of biological complexity driven by the natural selection of genetic and cultural information extends but is completely consistent with Schrödinger's physical argument because information-driven complexity is fundamentally an energetic phenomenon and not confined to biology. Nevertheless, the current increase in the overall complexity of biological organisation is also in large part a consequence of Lamarckian, rather than purely genetic, evolution. Perhaps it is an overdue vindication of 'le fondateur de la doctrine de l'évolution' – the founder of evolutionary doctrine (Figure 8.2).

9

Why DNA

and Not RNA?

Variation, in fact, is evolution.
—William Bateson, 1894.

Genetic Diversity and Complexity

In the latter part of the eighteenth century William Bateson (1894) asserted as a central tenet that 'Variation, in fact, is evolution'. By conflating the then novel science of genetics with established studies of biological variation – notably those of John Stevens Henslow (the first director of Cambridge University Botanic Garden and likely 'the father of variation') and his better-known student, Charles Darwin, as well as those of Alfred Wallace – Bateson argued that variation, whatever its source, is a central driver of evolution and for biological evolution genetics – and hence DNA – plays a major role.

One source of variation is genomic diversity within a single organism. Thus in lichens – 'simple' complex organisms – the acquisition of a complex organisation is associated with greater genetic diversity mediated by multiple genomes. In others, especially animals, epigenetic diversity plays a major role. But underpinning these mechanisms is the essential process of variation of genomic information – whether mediated by differential assortment during, for example, meiosis, or by changes in the base sequence itself. Mutability is key, but only within acceptable bounds. And it is just this mutability of the DNA sequence that is the engine for variation and thus the generation of genomic diversity. DNA fulfills five essential properties of an information carrier – it encodes information precisely, that information can be stored in a remarkably small volume, that information can be readily accessed, that information can be more or less precisely replicated and shuffled and most importantly from the perspective of evolution, that information is editable. Over the vast majority of biological evolutionary time DNA was the dominant vehicle encoding genetic information but by selecting for information, the resultant complex biological system evolved new forms of information processing and storage, further increasing the complexity of the system. Crucially the principles of Darwinian evolutionary variation and consequent adaptation apply both to DNA and novel information-processing systems. It is precisely this increasing sophistication of information processing in biology that is an important driver of evolution and complexity. Or in Boltzmann's words, '. . . all salvation for philosophy may be expected to come from Darwin's theory'.

Why DNA: And Not RNA?

Why DNA – and not RNA? Even now – more than 60 years after the discovery of its double-helical structure, the study of the diversity of structures assumed by this polymer is still revealing a plethora of biological roles that enhance and fine-tune its primary role as an information store.

Schrödinger's immense contribution was essentially to link energy, information and complexity – all physical concepts – as the defining characteristics of life. More information permits greater complexity, and the more complex a system, on average, the more energy is required for its maintenance. Most importantly, Schrödinger also recognised that biological information is mutable and so can change in subtle – and sometimes not so subtle – ways from generation to generation. It is this mutability that, by modulating the informational instructions in an organism's genome, provides the basis for Darwinian evolution by natural selection. Here again the key concept, probability, is essentially a physical parameter. Thus the probability of a single base change in, for example, a bacterial genome, resulting in a small increase in energy efficiency in the context of its immediate complex environment, is enormously greater than that of the de novo generation of the corresponding 'improved' bacterium. In other words, Darwinian evolution depends on and acts on pre-existing complexity – a superb example of Bayesian logic. There is consequently no mystery about the progressive and observed increase in biological complexity. It is simply a physical phenomenon – directional because it depends on previous evolution but in no way directed.

But because biological complexity is simply an alternative description of a particular complex chemical system in which information flux plays a dominant role, its efficient operation, just like that of a computer where information flux is also paramount, is subject to strict physical limits. While the mutability of the DNA information store is a crucial component of the burgeoning nature of biological complexity, it is the nucleic acids themselves – both DNA and RNA – that potentially determine the physical limits of the complex system we know as life. In biological systems a major function of DNA – shared with RNA – is to match up a sequence of complementary base-pairs to form a double helix. This is the underlying mechanism of digital information transfer. Arguably without this unique property, biology as we know it would not exist. The replication of genetic information, the transcription of selected genes and the translation of mRNA, or, in other words, the fundamentals of digital information flux within a cell, all depend on this one property. Crucially all these informational interactions are reversible – they occur within a narrow energetic window where the double helix can melt and then reanneal, or vice versa. In this respect a loss of the ability of a nucleic acid sequence to base-pair with its complement – for example, if the temperature were too high – would preclude information transfer and be inimical to the operation of the complex chemical system that we call 'life'. This transition from reversibility to effective irreversibility thus likely represents the upper energetic constraint of the biological complex system, above which the system would essentially become more disorganised or chaotic. And indeed evolution has provided many mechanisms that could guard against such a catastrophe, which could potentially be induced by unpredictable external energetic fluctuations. One such is the circular form of the small genomes of DNA plasmids and of prokaryotic organisms (see Figure 3.13). In an intact DNA circle even if the double helix completely denatures, the two strands remain closely intertwined and because of this renature very rapidly when the external conditions return to a more amenable milieu. Indeed, this principle is the practical basis for separating circular plasmid DNA from linear DNA in the laboratory. Likewise linear DNA domains, in which the DNA is fastened to a fixed point at each end, could perform a similar function. Another mechanism is the stabilisation of different energetic forms of DNA by DNA binding proteins – histones in eukaryotes and Archaea and nucleoid-associated proteins in bacteria. Such proteins in a complex with DNA can raise its melting temperature. Yet again, in thermophilic Archaea and bacteria the reverse gyrase enzyme uses energy to reduce negative superhelicity and hence increase the average melting temperature of circular DNA.

But if irreversible melting delineates the upper bound of the biological system, what determines the lower bound? Here again the principle of enabling information flux is likely paramount. In bacteria one response to lower temperatures is the adoption of an alternative base-pairing configuration in certain mRNAs facilitating their translation under otherwise less favourable conditions. In general information flux is favoured by dynamic, yet organised, milieux and consequently disfavoured by the formation of less dynamic condensed DNA-containing aggregates and crystalline DNA. Some of the more extreme examples of transcriptionally inactive packaged DNA are found in both certain bacterial and archaeal DNA viruses (DiMaio et al., 2015) and also in bacterial spores (Lee, 2008). In the former DNA is present as crystalline A-DNA, while in bacterial spores the DNA is stabilised as an A-like, although not canonical A-form, DNA. In this context sperm can be regarded as rather sophisticated DNA virus particles with transcriptionally inactive, highly condensed DNA. As in bacterial spores, the chromatin structure is stabilised by interactions with basic proteins or, in vertebrates, basic peptides such as protamine.

A crucial difference between DNA and RNA is that whereas many of the functions of RNA – for example in the ribosome, in transfer RNA and in mRNA – depend on the formation of a secondary structure often involving the formation of several, or even many, short double helices in a single RNA strand, for DNA the double-stranded structure is essential for its core functions of the conservation and stability of genomic

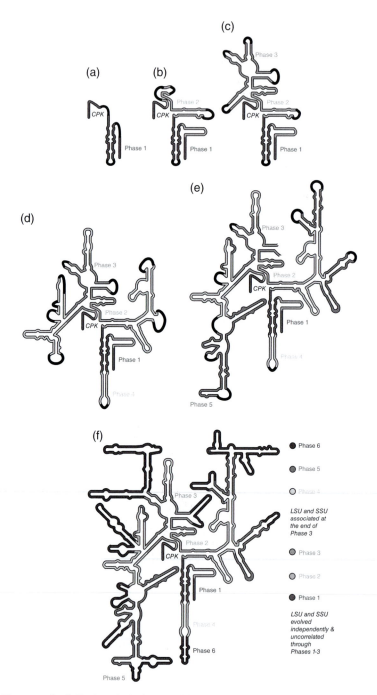

Figure 9.1 (a-e) The hypothesised sequential evolution of the RNA in the small ribosomal subunit from simple short RNA molecules existing in the primeval RNA world to the present-day

information as well as for the selection of that information during transcription. Of course, for some RNA viruses replication proceeds via a double-stranded intermediate, but in general such RNA genomes are very much shorter than DNA genomes, or even single DNA chromosomes, and so encode less total information. The functional distinction between RNA and DNA highlights a potential conflict to form, as in RNA, multiple double helices in a single strand and, as in DNA, a more stable continuous single double helix. Although the sequences of a particular RNA molecule and its cognate DNA region are necessarily the same, this conflict would likely be exacerbated with longer RNA molecules, if only because the selection pressures on the same sequences in DNA and RNA would likely differ.

The evolutionary progression is superbly exemplified by the longer RNA molecules in the large and small ribosomal subunits. Both these molecules, respectively termed 23S and 16S, adopt highly folded secondary structures that determine the structural integrity of the ribosomal subunits (Figure 9.1). Molecular archaeology has recently revealed that the current organisation of these molecules arose by the insertion of sequence blocks resulting in a progressive accretion that extended the structural complexity and versatility of the RNA molecules. The structural complexity reflects function and is selectable, but there is a likely cost. With a longer RNA molecule sequence integrity is compromised, and in a wholly double-stranded RNA form, replication efficiency might also be slowed by competition between competing double-helical configurations of the sequence. But with DNA both these costs can be mitigated. A continuous double-stranded structure would by conserving complementary information be compatible with the development of dedicated repair systems. Similarly the transcription of a single-stranded RNA molecule from DNA would facilitate its rapid folding into a functional structure. But when in the evolutionary process would DNA have appeared? The addition of proteins to the structure coincided with a

Figure 9.1 (*cont.*) association of much longer RNA molecules with proteins. The evolutionary stages are colour-coded to correspond with successive sequence insertions. *A black-and-white version of this figure will appear in some formats. For the colour version, refer to the plate section.*
Source: Adapted from Petrov et al. (2015) Proceedings of the National Academy of Science, USA, 112, 15396–15401.

substantial increase in the length of the rRNA molecules, while the
protein requirement would itself further increase the number of
genes and hence the information required for ribosome assembly.
In the preceding scenario where a ribosomal RNA increases in length
and completely any correlation between an RNA world and an RNA
plus DNA world cannot be identified.

But the question as to why RNA failed to retain its position as the
predominant genetic material remains, Although it is now almost
universally acknowledged that RNA represents the genesis of nucleic
acid mediated information transfer, the reason for subsequent evolutionary supremacy of DNA as the dominant genetic material is less
immediately obvious. A genetic information system containing both
DNA and RNA is in itself more complex than one with RNA alone, so
the crucial question to address is the nature of the evolutionary driver
for this increase in complexity. The most obvious differences between
RNA and DNA are both structural and functional. Today cellular
RNA molecules are overwhelmingly single-stranded – as mRNA,
tRNA, ribosomal RNA and many viral RNA genomes. There are a
few exceptions – some viral RNA genomes are double stranded. For
DNA the pattern is the exact opposite. It is overwhelmingly double
stranded except for a few single-stranded viral genomes such as that of
bacteriophage ΦX174. This structural distinction underlies a profound
functional difference in information content and transfer. Whereas
information transfer mediated by an RNA molecule is essentially
digital in nature, that by DNA is dual – it can utilise both digital and
analogue modes. Or more prosaically one molecule can act both as a
Babbage calculator and a computer chip. Both codon–anticodon and
DNA–RNA interactions depend on digital sequence recognition mediated by accurate base-pairing, as also does the reannealing of a DNA
duplex. In contrast the analogue information represented by DNA
helical repeat is a property encoded by a duplex and not by a single
nucleic acid strand. Base-pairing provides a structural basis for the
helical duplex but, to a first approximation apart from a limited
variation dependent on base composition, is not an active participant
in the variation of the helical repeat. Importantly this parameter is
uniquely sensitive to environmental conditions – it is affected by
temperature, DNA supercoiling and the ionic milieux – in effect, it
can potentially act as an environmental sensor (Figure 9.2).

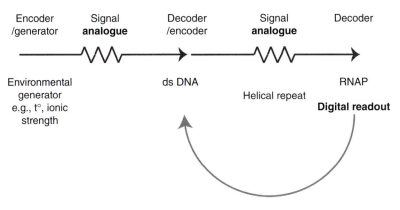

| Encoder /generator | Signal **analogue** | Decoder /encoder | Signal **analogue** | Decoder |

Environmental generator e.g., t°, ionic strength

ds DNA

Helical repeat

RNAP

Digital readout

NAPs, histones, etc

Figure 9.2 Double-stranded DNA – a convertor of analogue information. Expression of genetic information via the sensing of the helical repeat by, for example, RNA polymerase (RNAP) can create an informational loop, which in turn can modify as necessary the helical repeat and thus gene expression. The terms *encoder*, *signal* and *decoder* refer to the process of Shannon information transfer (see Figure 2.2). NAPs are bacterial nucleoid-associated proteins that perform related functions to eukaryotic histones.

Changes in the helical repeat have important consequences for processes that involve DNA melting. A simple example is provided by the phenomenon of salt shock in bacteria. When bacteria are transferred to a medium of high osmolarity, there is rapid influx of potassium ions raising the internal ionic strength. This is immediately accompanied by a sharp rise in energy availability from ATP and a consequent increase in the negative superhelical density of DNA (McClellan, Boublíková, Paleček, & Lilley, 1990; Hsieh, Rouvière-Yaniv, & Drlica, 1991). This is essentially a classic homeostatic response to a situation where the rise in internal ionic strength decreases the helical repeat (overwinds) the DNA. The increased energy availability activates the topoisomerase DNA gyrase, which in turn leads to increased negative superhelicity. This both increases the helical repeat of unbound DNA and provides torque for facilitating the melting of, say, promoter DNA in a polymerase-promoter DNA complex. A similar effect occurs when a non-growing (stationary) population of bacteria is transferred to fresh medium. Again an influx of potassium ions is accompanied by a higher negative superhelical density (Travers & Muskhelishvili, 2020). Another simple example is the effect of temperature on DNA supercoiling in bacteria (Goldstein & Drlica, 1984). Plasmid DNA isolated from bacteria growing at higher temperatures exhibit a less negative linking number than at lower temperatures. In other words, temperature dependent untwisting is compensated by lower levels of negative superhelicity. This effect is comparable to the introduction of positive supercoils in the DNA of hyperthermophilic bacteria growing at high temperatures. Again this reduces the DNA untwisting that would normally occur in, say, the vicinity of an undersea volcanic vent, to levels where the double helix remains intact. This general mode of regulation acts on a single physical parameter of DNA – the helical repeat. The plethora of varied protein structures lacks such a common feature. In this case adaptation to growth at high temperatures often minimises the occurrence of flexible extrusions from the protein surface, so effectively reducing the entropy associated with the whole molecule and inadvertently providing a boon for protein crystallographers.

The observation that the transition from an RNA world to an RNA plus DNA world is accompanied by a step change in information capacity is consistent with the premise that an increase in complexity is linked to an increase in information potential (Figure 9.3). DNA is the

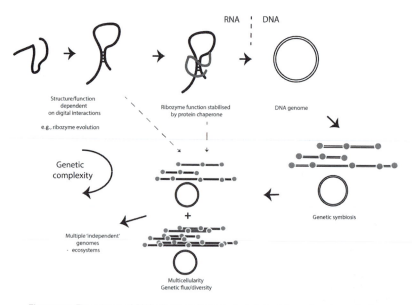

Figure 9.3 The advent of DNA allows an increase in cellular informational complexity.

vehicle for this increased information capacity, but the evolutionary driver for the change – always with the caveat that sufficient energy is available – is likely information itself. For the RNA to RNA plus DNA, transition was not the only change, and arguably not even the most important, in information-processing capacity during biological evolution. It was likely preceded by the adoption of a mechanism for converting the 4-letter digital code in RNA to a more complex 20-letter code in protein sequences (Chapter 2). This enabled the appearance of more precisely defined macromolecules with potentially greater specificity for enzymatic catalysis accompanied by an increase in overall organisation in accord with Schrödinger's negentropy hypothesis. A crucial role of DNA would be to store this information in an accessible and modular form. These changes in information manipulation would also facilitate the rise of self-referentiality. It follows that Darwinian selection ultimately acts on the expression of available information, and not all of this information is necessarily encoded by DNA. The increased sophistication of information processing would be accompanied by, and perhaps dependent on, an increased sophistication of mechanisms for energy harnessing. Ultimately (biological) complexity is dependent on available information and available information is dependent on available energy. Or, again in Boltzmann's words: 'Available energy is the main object at stake in the struggle for existence and the evolution of the world'.

General Reading and Bibliography

General Reading

(Not all the following are directly cited but, depending on the level of understanding wished for, the reader should find that all are well worth reading for a thorough appreciation of the role of DNA in biological evolution.)

Bonner, J. (1988) *The evolution of complexity by means of natural selection.* Princeton University Press, Princeton, NJ.

Calladine, C.R., Drew H.R., Luisi, B.F, Travers, A.A. (2004) *Understanding DNA and how it works.* Elsevier, San Francisco, CA.

Coyne, J.A. (2009) *Why evolution is true.* Oxford University Press, Oxford, UK.

(A very similar argument to Dawkins' book, 'The greatest show on Earth', but in a somewhat less polemic style.)

Darwin, C. (1859) *On the origin of species by means of natural selection, or the preservation of favoured races in the struggle for life.* John Murray, London, UK.

(A modern paperback edition of this classic was published by Oxford University Press with a penetrating introduction by Gillian Beer in 1996.)

Dawkins, R. (1976) *The selfish gene.* Oxford University Press, Oxford, UK.

(1982) *The extended phenotype.* Oxford University Press, Oxford, UK.

(2004) *The ancestor's tale.* Weidenfeld and Nicholson, London, UK.

(2009) *The greatest show on Earth.* Bantam Press, London, UK.

Greaves, M. (2000) *Cancer. The evolutionary legacy.* Oxford University Press, Oxford, UK.

Jencks, C. (1995) *The architecture of the jumping universe.* Academy Editions, London.

(A very personal view of complexity – akin to that of Chaisson.)

Jones, S. (1999) *Almost like a whale.* Transworld Publishers, London, UK.

Jukes, T.H. (1963) Observations on the possible nature of the genetic code. *Biochemical and Biophysical Research Communications* 10, 155–159.

Kauffman, S.A. (1993) *The origins of order.* Oxford University Press, Oxford, UK.

Koonin, E.V. (2012) *The logic of chance: The nature and origin of biological evolution.* Pearson Education Inc., Upper Saddle River, NJ.

Lane, N. (2015) *The vital question.* Profile Books, London, UK.

Lineweaver, C., Davies, P., Ruse, M. (Eds.) (2013) *Complexity and the arrow of time.* Cambridge University Press, Cambridge, UK.

Mayr, E. (1963) *Animal species and evolution.* Belknap Press, Cambridge, MA.

Muskhelishvili, G. (2015) *DNA information: Laws of perception.* Springer, Dordrecht, Germany.

Papineau, D. (2018) Thomas Bayes and the crisis in science. *Times Literary Supplement,* 28 June.

Renfrew, C. (1987) *Archaeology and language: The puzzle of Indo-European origins,* Pimlico, London, UK.

Schrödinger, E. (1944) *What is life?* Cambridge University Press, Cambridge, UK.

Smith, E., Morowitz, H.J. (2016) *The origin and nature of life on earth.* Cambridge University Press, Cambridge, UK.

Tainter, J.A. (1988) *The collapse of complex societies.* Cambridge University Press, Cambridge, UK.

Vologodskii A. (2015) *Biophysics of DNA.* Cambridge University Press, Cambridge, UK, 2015.

von Neumann, J. (1958) *The computer and the brain.* Yale University Press, New Haven, CT.

Wilson, E.O. (1992) *The diversity of life.* Allen Lane, London, UK.

Heading Quotes

Chapter 1

Boltzmann, L. (1886) *Der zweite Hauptsatz der mechanischen Warmetheorie.* Vienna: Gerold, p. 21.

Lewis, G.N. (1930) From a letter to Irving Langmuir, 5 August 1930. Quoted in Reingold, N. (1981) *Science in America: A documentary history 1900–1939.* University of Chicago Press, Chicago, IL, 400.

Chapter 2

Landauer, R. (1991) The physical nature of information. *Physics Letters,* 217, 188–193.

Maxwell, J.C. (1850) In Maxwell, J.C., Harman, P.M. (Eds.) (1990) *The scientific letters and papers of James Clerk Maxwell, Vol. 1, 1846–1862,* p. 197. Cambridge University Press, Cambridge, UK.

Monod, J. (1970) *Le Hasard et la Nécessité: Essai sur la philosophie naturelle de la biologie moderne.* Paris: Éditions du Seuil.

Szilard, L. (1929) In Über die Entropieverminderung in einem thermodynamischen System bei Eingriffen intelligenter Wesen (On the reduction of entropy in a thermodynamic system by the intervention of intelligent beings). *Zeitschrift für Physik* 53, 840–856.

Chapter 3

Monod, J. (1970) *Le Hasard et la Nécessité: Essai sur la philosophie naturelle de la biologie moderne.* Paris: Éditions du Seuil.

Chapter 4

Boltzmann, L. (1905) *Populäre Schriften, 1905.* Leipzig: Verlag von Johann Ambrosius Barth, p. 396.

Luhmann, N. (1984) *Soziale Systeme: Grundriß einer allegemeinen Theorie*. Surkamp Verlag, Frankfurt-am-Main, Germany.
English edition: translated by Bednarz, J., Jr. (1995) *Social systems*, Stanford University Press, Stanford, CA.

Chapter 5

Dawkins, R. (1982) *The extended phenotype*. Oxford University Press, Oxford, UK.
Mayr, E. (1963) *Animal species and evolution*. Belknap Press, Cambridge, MA.

Chapter 7

Darwin, C. (1871) From a letter to J.D. Hooker, 1 February 1871.

Chapter 8

Mayr, E. (1963) *Animal species and evolution*. Belknap Press, Cambridge, MA.

Chapter 9

Bateson, W. (1894) *Materials for the study of variation treated with especial regard to discontinuity in the origin of species*. Macmillan, London, UK.

Citations

(The literature covering the topics discussed is truly immense. I apologise for my inadvertent omissions.)

Adam, P.S., Borrel, G., Gribaldo, S. (2018) Evolutionary history of carbon monoxide dehydrogenase/acetyl-CoA synthase, one of the oldest enzymatic complexes. *Proceedings of the National Academy of Sciences of the USA* 115, E1166–E1173.

(2019) An archaeal origin of the Wood–Ljungdahl H₄MPT branch and the emergence of bacterial methylotrophy. *Nature Microbiology* 4, 2155–2163.

Adami, C. (2002) What is complexity? *Bioessays* 24, 12, 1085–1094.

Adami, C., Ofria, C., Collier, T.C. (2000) Evolution of biological complexity. *Proceedings of the National Academy of Sciences of the USA* 97, 9, 4463–4468.

Allman, S., Baldwin, I.T. (2010) Insects betray themselves in nature to predators by the rapid isomerization of green leaf volatiles. *Science* 329, 1075–1078.

Amouyal, M., Buc, H. (1987) Topological unwinding at strong and weak promoters by RNA polymerase. A comparison between the lac wild type and the UV5 sites of *Escherichia coli. Journal of Molecular Biology* 195, 795–808.

Anderson, P., Bauer, W. (1978) Supercoiling in closed circular DNA: Dependence upon ion type and concentration. *Biochemistry* 17, 594–601.

Aragão, L.E.O.C. (2012) The rainforest's water pump. *Nature* 489, 217–218.

Arnott, S. (2006) Historical article: DNA polymorphism and the early history of the double helix. *Trends in Biochemical Sciences* 31, 349–354.

Avery, O.T., Macleod, C.M., McCarty, M. (1944) Studies on the chemical nature of the substance inducing transformation of pneumococcal types: Induction of transformation by a desoxyribonucleic acid fraction isolated from pneumococcus type III. *Journal of Experimental Medicine.* 79, 137–158.

Baldwin, G.S., Brooks, N.J., Robson, R.E., Wynveen, A., Goldar, A., Leikin, S., Seddon, J.M., Kornyshev, A.A. (2008) DNA double helices recognize mutual sequence homology in a protein free environment. *The Journal of Physical Chemistry B* 112, 1060–1064.

Barbieri, M. (2016) What is information? *Philosophical Transactions of the Royal Society A: Mathematical, Physical and Engineering Sciences* 374, pii, 20150060.

Barrangou, R., Fremaux, C., Deveau, H., Richards, M., Boyaval, P., Moineau, S., Romero, D.A., Horvath, P. (2007) CRISPR provides acquired resistance against viruses in prokaryotes. *Science* 315, 1709–1712.

Bateson, W. (1894) *Materials for the study of variation treated with especial regard to discontinuity in the origin of species.* Macmillan, London, UK.

Bayes, T. (1763) An essay toward solving a problem of the doctrine of chances. *Philosophical Transactions*, 307–418.

Bérut, A., Arakelyan, A., Petrosyan, A., Ciliberto, S., Dillenschneider, R., Lutz, E. (2012) Experimental verification of Landauer's principle linking information and thermodynamics. *Nature* 483, 187–189.

Bird, A. (2007) Perceptions of epigenetics. *Nature*, 447, 396–398.

Boehnke, P., Bell, E.A., Stephan, T., Trappitsch, R., Keller, C.B., Pardo, O.S., Davis, A.M., Harrison, T.M., Pellin, M.J. (2018) Potassic, high-silica Hadean crust. *Proceedings of the National Academy of Sciences of the USA* 115, 6353–6356.

Böhle, U.-R., Hilger, H.H., Martin, W.F. (1996) Island colonization and evolution of the insular woody habit in Echium L. (Boraginaceae). *Proceedings of the National Academy of Sciences of the USA* 93, 11740–11745.

Brillouin, L. (1953) Negentropy principle of information. *Journal of Applied Physics* 24, 1152–1163.

(1956) *Science and information theory.* Academic Press, New York, NY.

Brouns, S.J., Jore, M.M., Lundgren, M., Westra, E.R., Slijkhuis, R.J., Snijders, A.P., Dickman, M.J., Makarova, K.S., Koonin, E.V., van der Oost, J. (2008) Small CRISPR RNAs guide antiviral defense in prokaryotes. *Science* 321, 960–964.

Buchner, P. (1965) *Endosymbiosis of animals with plant microorganisms* (English translation by Bertha Müller). Interscience, New York, NY.

Cech, T.R., Zaug, A.J., Grabowski, P.J. (1981) In vitro splicing of a ribosomal RNA of Tetrahymena: Involvement of a guanosine nucleotide in the excision of the intervening sequence. *Cell* 27, 487–496.

Chaisson, E.J. (2011) Energy rate density as a complexity metric and evolutionary driver. *Complexity* 16, 27–40.

(2013) Using complexity science to search for unity in the natural sciences. In *Complexity and the arrow of time* (ed. Lineweaver, C., Davies, P., Ruse, M.). Cambridge University Press, Cambridge, UK.

(2015) Energy flows in low-entropy complex systems. *Entropy* 17, 8007–8018.

Church, R., McCarthy, B.J. (1967) Changes in nuclear and cytoplasmic RNA in regenerating mouse liver. *Proceedings of the National Academy of Sciences of the USA* 58, 1548–1555.

Ciciriello, F., Costanzo, G., Pino, S., Crestini, C., Saladino, R. & Di Mauro, E. (2008) Molecular complexity favors the evolution of ribopolymers. *Biochemistry* 47, 2732–2742.

Cleveland, L.R., Grimstone, A.V. (1964) The fine structure of the flagellate *Mixotricha paradoxa* and its associated micro-organisms. *Proceedings of the Royal Society B: Biological Sciences*, 159, 668–686.

Cobb, M. (2017) 60 years ago Francis Crick changed the logic of biology. *PLoS Biology* 15(9), e2003243.

Coen, E. (2001) Goethe and the ABC model of flower development. *Comptes Rendus de l'Académie des Sciences – Series III* 324, 523–530.

Cozzarelli, N.R., Boles, T.C., White, J.H. (1990) Primer on the topology and geometry of DNA supercoiling. In *DNA topology and its biological effects* (eds. Cozzarelli, N.R., Wang, J.C.), pp. 139–184. Cold Spring Harbor Laboratory Press, Cold Spring Harbor, NY.

Crawford, J.L., Kolpak, F.J., Wang, A.H., Quigley, G.J., van Boom, J.H., van der Marel, G., Rich, A. (1980) The tetramer d(CpGpCpGp) crystallizes as a left-handed double helix. *Proceedings of the National Academy of Sciences of the USA* 77, 4016–4020.

Crick, F.H.C. (1958) On protein synthesis. *Symposia of the Society for Experimental Biology* 12, 138–163.

(1966) Codon-anticodon pairing: the wobble hypothesis. *Journal of Molecular Biology* 19, 548–555.

(1968) The origin of the genetic code. *Journal of Molecular Biology* 38, 367–379.

Cubas, P., Vincent, C., Coen, E. (1999) An epigenetic mutation responsible for natural variation in floral symmetry. *Nature* 401, 157–161 (1999).

Danchin, A., Nikel, P.I. (2019) Why nature chose potassium. *Journal of Molecular Evolution* 87, 271–288.

Darwin, C. (1842) *Essays of 1842 and 1844.* http://darwin-online.org.uk/converted/pdf/1909_Foundations_F1556.pdf

Dawkins, R. (1976) *The selfish gene.* Oxford University Press, Oxford, UK.

de Frutos, M., Leforestier, A., Livolant, F. (2014) Relationship between the genome packing in the bacteriophage capsid and the kinetics of DNA ejection. *Biophysical Reviews and Letters* 9, 81–104.

Dibrova, D.V., Galperin, M.Y., Koonin, E.V., Mulkidjanian, A.Y. (2015) Ancient systems of sodium/potassium homeostasis as precursors of membrane energetics. *Biochemistry (Moscow)* 80, 495–516.

DiMaio, F., Yu, X., Rensen, E., Krupovic, M., Prangishvili, D., Egelman, E.H. (2015) A virus that infects a hyperthermophile encapsidates A-form DNA. *Science* 348, 914–917.

Doolittle, W.F., Sapienza, C. (1980) Selfish genes, the phenotype paradigm and genome evolution. *Nature* 284, 601–603.

Eddington, A.S. (1928) Quote from *The Nature of the Physical World.* Cambridge University Press, Cambridge, UK, 1928.

El Hassan, M.A., Calladine, C.R. (1997) Conformational characteristics of DNA: Empirical classifications and a hypothesis for the conformational behaviour of dinucleotide steps. *Philosophical Transactions of the Royal Society A:* 355, 43–100.

Francis, R., Read, D.J. (1984) Direct transfer of carbon between plants connected by vesicular arbuscular mycorrhizal mycelium. *Nature* 307, 53–56.

Gilbert, N., Allan, J. (2014) Supercoiling in DNA and chromatin. *Current Opinion in Genetics and Development* 25, 15–21.

Glaab, F., Kellermeier, M., Kunz, W., Morallon, M., Garciá-Ruiz. (2012) Formation and evolution of chemical gradients and potential differences across self-assembling inorganic membranes. *Angewandte Chemie International Edition* 51, 4317–4321.

Goldstein, E., Drlica, K. (1984) Regulation of bacterial DNA supercoiling: Plasmid linking numbers vary with growth temperature. *Proceedings of the National Academy of Sciences of the USA* 81, 4046–4050.

Gorin, A.A., Zhurkin, V.B., Olson, W.K. (1995) B-DNA twisting correlates with base-pair morphology. *Journal of Molecular Biology* 247, 34–48.

Greaves, M. (2000) *Cancer. The evolutionary legacy*. Oxford University Press, Oxford, UK.

Guerrier-Takada, C., Gardiner, K., Marsh, T., Pace, N, Altman, S. (1983) The RNA moiety is the catalytic subunit of the enzyme. *Cell* 35, 849–857.

Hamilton, W. (1964) The genetical evolution of social behaviour. *Journal of Theoretical Biology* 7, 1–16, 17–52.

Hemmo, M., Shenker, O.R. (2012) *The road to Maxwell's demon*. Cambridge University Press, Cambridge, UK.

Holland, T. (2005) *Persian fire: The first world empire and the battle for the west*. Random House, New York, NY.

Hsieh, L.S., Rouvière-Yaniv, J., Drlica, K. (1991) Bacterial DNA supercoiling and [ATP]/[ADP] ratio: Changes associated with salt shock. *Journal of Bacteriology* 173, 3914–3917.

Hunter, C.A. (1993) Sequence-dependent DNA structure. The role of base stacking interactions. *Journal of Molecular Biology* 230, 1025–1054.

(1996) Sequence-dependent DNA structure. *Bioessays* 18, 157–162.

Huxley, J. (1959) Introduction. In de Chardin, P.T. *The phenomenon of man*. Harper, New York, NY.

Hyman, A.A., Weber, C.A., Jülicher, F. (2014) Liquid-liquid phase separation in biology. *Annual Review of Cell and Developmental Biology* 30, 39–58.

Inoue, S., Sugiyama, S., Travers, A.A., Ohyama, T. (2007) Self-assembly of double-stranded DNA molecules at nanomolar concentrations. *Biochemistry* 46, 164–171.

Jaynes, E.T. (2003) *Probability theory: The logic of science*. Cambridge University Press, Cambridge, UK. See Chapter 22 for quote.

Johnson, R.E., Prakash, L., Prakash, S. (2005) Biochemical evidence for the requirement of Hoogsteen base pairing for replication by human DNA polymerase I. *Proceedings of the National Academy of Sciences of the USA* 102, 10466–10471.

Kato, J., Misra, T.K., Chakrabarty, A.M. (1990) AlgR3, a protein resembling eukaryotic histone H1, regulates alginate synthesis in *Pseudomonas aeruginosa*. *Proceedings of the National Academy of Sciences of the USA* 87, 2887–2891.

Koonin, E.V. (2009) Darwinian evolution in the light of genomics. *Nucleic Acids Research* 37, 111–1034.

Koonin, E.V., Wolf, Y.I. (2009) Is evolution Darwinian and/or Lamarckian? *Biology Direct* 4, 42.

Koski, J.V., Maisi, V.F., Pekola. J.P., Averin, D.V. (2014) Experimental realization of a Szilard engine with a single electron. *Proceedings of the National Academy of Sciences of the USA* 111, 13786–13789.

Lane, N., Allen, J.F., Martin, W. (2010) How did LUCA make a living? Chemiosmosis in the origin of life. *Bioessays* 32(4), 271–280.

Larson, A.G., Elnatan, D., Keenen, M.M., Trnka, M.J., Johnston, J.B., Burlingame, A.L., Agard, D.A., Redding, S., Narlikar, G.J. (2017) Liquid droplet formation by HP1α suggests a role for phase separation in heterochromatin. *Nature* 547, 236–240.

Laundon, C.H., Griffith, J.D. (1988). Curved helix segments can uniquely orient the topology of supertwisted DNA. *Cell* 52, 545–549.

Lazcano, A. (2010) Historical development of origins research. *Cold Spring Harbor Perspectives in Biology* 2, a002089.

Lee, K.S., Bumbaca, D., Kosman, J., Setlow, P., Jedrzejas, M.J. (2008) Structure of a DNA-protein complex essential for protection in spores of Bacillus species. *Proceedings of the National Academy of Sciences of the USA* 105, 2806–2811.

Leforestier, A., Livolant, F. (2009). Structure of toroidal DNA collapsed inside the phage capsid. *Proceedings of the National Academy of Sciences of the USA* 106, 9157–9162.

(2010) The bacteriophage genome undergoes a succession of intracapsid phase transitions upon DNA ejection. *Journal of Molecular Biology* 396, 384–395.

Lewis, G.N. (1926) quote from pp. 158–159 of *The Anatomy of Science*. Yale University Press, New Haven, CT.

Lilley, D.M.J. (1980) The inverted repeat as a recognisable structural feature in supercoiled DNA molecules. *Proceedings of the National Academy of Sciences of the USA* 77, 6468–6472.

Linnaeus, C. (1749) *De Peloria*. Diss. Ac. Amoenitates Academicae III, Uppsala.

Liu, L.F., Wang, J.C. (1987) Supercoiling of the DNA template during transcription. *Proceedings of the National Academy of Sciences of the USA* 84, 7024–7027.

Lotka, A.J. (1922a) Contributions to energetics of evolution. *Proceedings of the National Academy of Sciences of the USA* 8, 147–151.

(1922b) Natural selection as a physical principle. *Proceedings of the National Academy of Sciences of the USA* 8, 151–154.

Luger, K., Mäder, A.W., Richmond, R.K., Sargent, D.F., Richmond, T.J. (1997) Crystal structure of the nucleosome core particle at 2.8 Å resolution. *Nature* 389, 251–260.

Macallum, A.B. (1926) The paleochemistry of the body fluids and tissues. *Physiological Reviews* 6, 316–357.

Mace, H.A., Pelham, H.R., Travers, A.A. (1983) Association of an S1 nuclease-sensitive structure with short direct repeats 5' of *Drosophila* heat shock genes. *Nature* 304, 555–557.

MacFarlane, R. (2019) *Underland.* Hamish Hamilton, London, UK.

Margulis, L. (1991) Symbiosis and symbioticism. In *Symbiosis as a source of evolutionary innovation: speciation and morphogenesis* (ed. Margulis, L.), pp. 1–13, MIT Press, Boston, MA.

Marr, C., Geertz, M., Hütt, M.T., Muskhelishvili, G. (2008) Dissecting the logical types of network control in gene expression profiles. *BMC Systems Biology.* 2, 18.

Martin, W., Kowallik, K.V. (1999) Annotated English translation of Mereschkowsky's 1905 paper 'Über Natur und Ursprung der Chromatophoren im Pflanzenreiche'. *European Journal of Phycology* 34, 287–295.

Maurer, S., Fritz, J., Muskhelishvili, G., Travers, A. (2006) RNA polymerase and an activator form discrete subcomplexes in a transcription initiation complex. *EMBO Journal* 25, 3784–3790.

Mayr, E. (1963) *Animal species and evolution.* Belknap Press, Cambridge, MA.

McClellan, J.A., Boublíková, P., Paleček, E., Lilley, D.M.J. (1990) Superhelical torsion in cellular DNA responds directly to environmental and genetic factors. *Proceedings of the National Academy of Sciences of the USA* 87, 8373–8377.

McClintock, B. (1950) The origin and behavior of mutable loci in maize. *Proceedings of the National Academy of Sciences of the USA* 36, 344–355.

Medvedkin, V.N., Permyakov, E.A., Klimenko, L.V., Mitin, Y.V., Matsushima, N., Nakayama, S., Kretsinger, R.H. (1995) Interactions of (Ala*Ala*Lys*Pro)n and (Lys*Lys*Ser*Pro)n with DNA. Proposed coiled-coil structure of AlgR3 and AlgP from Pseudomonas aeruginosa. *Protein Engineering, Design and Selection* 8, 63–70.

Melis, P. (2003) *The Nuragic civilization.* Carlo Delfino editore, Sassari, Italy.

Mereschkowsky, C. (1905). Über Natur und Ursprung der Chromatophoren im Pflanzenreiche. *Biologisches Centralblatt* 25, 593–604.

(1910). Theorie der zwei Plasmaarten als Grundlage der Symbiogenesis, einer neuen Lehre von der Entstehung der Organismen. *Biologisches Centralblatt* 30, 278–288, 289–303, 321–347, 353–367.

Miller, S.L. (1953) A production of amino acids under possible primitive Earth conditions. *Science* 117, 528–529.

Miller, S.L., Urey, H.C. (1959) Organic compound synthesis on the primitive earth. *Science* 130, 245–251.

Minyat, E.E., Khomyakova, E.B., Petrova, M.V., Zdobnov, E.M., Ivanov, V.I. (1995) Experimental evidence for slipped loop DNA, a novel folding type for polynucleotide chain. *Journal of Biomolecular Structural and Dynamics* 13, 523–527.

Mizuno, T., Chou, M.-Y., Inouye, M. (1984) A unique mechanism regulating gene expression: Translational inhibition by a complementary RNA transcript. *Proceedings of the National Academy of Sciences of the USA* 81, 1966–1970.

Molina, E.C. (1963) *Two papers by Bayes with commentaries.* Hafner Publishing Co., New York, NY.

Morowitz, H.J. (1978) *Foundations of bioenergetics.* Academic Press, New York, NY.

Moyroud, E., Wenzel, T., Middleton, R., Rudall, P.J., Banks, H., Reed, A., Mellers, G., Killoran, P., Westwood, M.M., Steiner, U., Vignolini, S., Glover, B.J. (2017) Disorder in convergent floral nanostructures enhances signalling to bees. *Nature* 550, 469–474.

Mulkidjanian, A.Y., Bychkov, A.Y., Dibrova, D.V., Galperin, M.Y., Koonin, E.V. (2012) Origin of first cells at terrestrial, anoxic geothermal fields. *Proceedings of the National Academy of Sciences of the USA* 109, E821–E830.

Muskhelishvili, G., Travers, A. (1997) The stabilisation of DNA microloops by FIS – a mechanism for torsional transmission in transcription activation and DNA inversion. *Nucleic Acids and Molecular Biology* 11, 179–190.

Nigatu, D., Henkel, W., Sobetzko, P., Muskhelishvili, G. (2016) Relationship between digital information and thermodynamic stability in bacterial genomes. *EURASIP Journal on Bioinformatics and Systems Biology*, 4.

Niklas, K.J. (1986) Large-scale changes in plant and animal terrestrial communities. In *Patterns and processes in the history of life* (eds. Raup, D.M., Jablonski, D.), pp. 383–405, Springer, Berlin, Germany.

Nowak, M.A., Tarnita, C.E., Wilson, E.O. (2010) The evolution of eusociality. *Nature* 466, 1057–1062.

Olson, W.K., Gorin, A.A., Lu, X.-J., Hock, L.M., Zhurkin, V.B. (1998) DNA sequence-dependent deformability deduced from protein-DNA crystal complexes. *Proceedings of the National Academy of Sciences of the USA* 95, 11163–11168.

Oparin, A.I. (1936) *Vozniknovenie zhizni na zemle.* Moscow: Izd. Akad. Nauk SSSR. (English translation with annotations by Morgulis, S., Oparin, A.I. (1938) *The origin of life.* New York: Macmillan.)

Orgel, L.E. (1968) Evolution of the genetic apparatus. *Journal of Molecular Biology* 38, 381–393.

Orgel, L.E., Crick, F.H.C. (1980) Selfish DNA: The ultimate parasite. *Nature*, 284 604–607.

Owen, D.J., Ornaghi, P., Yang, J.-C., Lowe, N., Evans, P.R., Ballario, P., Filetici, P., Travers, A.A. (2000) The structural basis for the recognition of acetylated histone H4 by the bromodomain of histone acetyltransferase Gcn5p. *EMBO Journal* 19, 6141–6149.

Parrondo, J.M.R., Horowitz, J.M., Sagawa, T. (2015) Thermodynamics of information. *Nature Physics* 11, 131–139.

Pearce, B.K.D., Pudritz, R.E., Semenov, D.A., Henning, T.K. (2017) Origin of the RNA world. The fate of nucleobases in warm little ponds. *Proceedings of the National Academy of Sciences of the USA* 114, 11327–11332.

Pennisi, E. (2013) The CRISPR craze. *Science* 341, 833–836.

(2016) A lichen ménage à trois. *Science* 353, 337.

Penrose, R. (2010) *Cycles of time.* The Bodley Head, London, UK.

Petrov, A.S., Gulen, B., Norris, A.M., Kovacs, N.A., Bernier, C.R, Lanier, K.A., Fox, G.E., Harvey, S.C., Wartell, R.M., Hud, N.V., Williams, L.D. (2015) History of the ribosome and the origin of translation. *Proceedings of the National Academy of Sciences of the USA* 112, 15396–15401.

Portier, P. (1918) *Les symbiotes.* Masson, Paris, France.

Protozanova, E., Yakovchuk, P., Frank-Kamenetskii, M. (2004) Stacked-unstacked equilibrium at the nick site of DNA. *Journal of Molecular Biology* 342, 775–785.

Raff, R.A., Kaufmann, T.C. (1983) *Embryos, genes and evolution.* Macmillan, New York, NY.

Reiter, N.J., Osterman, A., Torres-Larios, A., Swinger, K.K., Pan, T., Mondragón, A. (2010) Structure of a bacterial ribonuclease P holoenzyme in complex with tRNA. *Nature* 468, 784–789.

Renfrew, C. (1987) *Archaeology and language: The puzzle of Indo-European origins*, Pimlico, London, UK.

Rhodes, D., Klug, A. (1980) Helical periodicity of DNA determined by enzyme digestion. *Nature* 286, 573–578.

Roach, J.C., Glusman, G., Smit, A.F.A., Huff, C. D., Hubley, R., Shannon, P.T., Rowen, L., Pant, K.P., Goodman, N., Bamshad, M., Shendure, J., Drmanac, R., Jorde, L.B., Hood, L., Galas, D.J. (2010) Analysis of genetic inheritance in a family quartet by whole-genome sequencing. *Science* 328, 636–639.

Rosenberg, E., Koren, O., Reshef, L., Efrony, R., Zilber-Rosenberg, I. (2007) The role of microorganisms in coral health, disease and evolution. *Nature Reviews Microbiology* 5, 355–362.

Saladino, R., Botta, G., Bizzarri, B.M., Di Mauro, E., Garciá Ruiz, J.M. (2016) A global scale scenario for prebiotic chemistry: Silica-based self-assembled mineral structures and formamide. *Biochemistry* 55, 2806–2811.

Saladino, R., Di Mauro, E., Garciá-Ruiz, J.M. (2019) A universal geochemical scenario for formamide condensation and prebiotic chemistry. *Chemistry*, 25, 3181–3189.

Satchwell, S.C., Drew, H.R., Travers, A.A (1986) Sequence periodicities in chicken nucleosome core DNA. *Journal of Molecular Biology* 191, 659–675.

Schimper, A.F.W. (1883) Über die Entwicklung der Chorophyllkörner und Farbkörper. *Botanische Zeitung 41*, 105–114.

Schneider, T.D. (1997) Information content of individual genetic sequences. *Journal of Theoretical Biology* 189, 427–441.

Schwendener, S. (1867). Über die wahre Natur der Flechten. *Verhandlungen der Schweizerischen Naturforschenden Gesellschaft in Rheinfelden* 51, 88–90.

Shannon, C. (1948) A mathematical theory of communication. *Bell System Technical Journal* 27, 379–423.

Sharov, A.A. (2016) Coenzyme world model of the origin of life. *Biosystems* 144, 8–17.

Simard, S.W., Perry, D.A., Jones, M.D., Myrold, D.D., Durall, D.M., Molina, R. (1997) Net transfer of carbon between ectomycorrhizal tree species in the field. *Nature* 388, 579–582.

Simon, H.A. (1962) The architecture of complexity. *Proceedings of the American Philosophical Society* 106, 467–482.

Spracklen, D.V., Arnold, S.R., Taylor, C.M. (2012) Observations of increased tropical rainfall preceded by air passage over forests. *Nature* 482, 282–285.

Strom, A.R., Emelyanov, A.V., Mir, M., Fyodorov, D.V., Darzacq, X., Karpen, G.H. (2017) Phase separation drives heterochromatin domain formation. *Nature* 547, 241–245.

Sundquist, W.I., Klug, A. (1989) Telomeric DNA dimerizes by formation of guanine tetrads between hairpin loops. *Nature* 342, 825–829.

Sutherland, J.D. (2016) The origin of life – out of the blue. *Angewandte Chemie International Edition* 55, 104–121.

Tainter, J.A. (1988) *The collapse of complex societies*. Cambridge University Press, Cambridge, UK.

Thomson, W. (Lord Kelvin) (1879) The sorting demon of Maxwell. *Nature* 20, 126.

Toyabe, S., Sagawa, T., Ueda, M., Muneyuki, E., Sano, M. (2010) Experimental demonstration of information-to-energy conversion and validation of the generalized Jarzynski equality. *Nature Physics* 6, 988–992.

Travers, A. (2006) The evolution of the genetic code revisited. *Origins of Life and Evolution of Biospheres* 36, 549–555.

Travers, A., Muskhelishvili, G. (2020) Chromosomal organisation and regulation of genetic function in *Escherichia coli* integrates the DNA analog and digital information. *EcoSal Plus* 9(1). doi: 10.1128/ecosalplus.ESP-0016-2019.

Tuovinen, V., Ekman, S., Thor, G., Vanderpool, D., Spribille, T., Johannesson, H. (2019) Two basidiomycete fungi in the cortex of wolf lichens. *Current Biology* 29, 476–483.

Turner, A.L., Watson, M., Wilkins, O.G., Cato, L., Travers, A., Thomas, J.O., Stott, K. (2018) Highly disordered histone H1-DNA model complexes and their condensates. *Proceedings of the National Academy of Sciences of the USA* 115, 11964–11969.

Van Roey, K., Uyar, B., Davey, N. (2014) Short linear motifs: Ubiquitous and functionally diverse protein interaction modules directing cell regulation. *Chemical Reviews* 114, 6733–6778.

Wang, J.C. (1979) Helical repeat of DNA in solution. *Proceedings of the National Academy of Sciences of the USA* 76, 200–203.

Watson, J.D., Crick, F.H.C. (1953) Molecular structure of nucleic acids: A structure for deoxyribonucleic acid. *Nature* 171, 737–738.

Watson, M., Stott, K. (2019) Disordered domains in chromatin binding proteins. *Essays in Biochemistry* 63, 147–156.

Weiss, M.C., Sousa, F.L, Mrnjavac, N., Neukirchen, S., Roettger, M., Nelson-Sathi, S., Martin, W.F. (2016). The physiology and habitat of the last universal common ancestor. *Nature Microbiology* 1(9), 16116.

Williams, T.A., Szöllősi, G.J., Spang, A., Foster, P.G., Heaps, S.E., Boussau, B., Ettema, T.J.G., Embley, T.M. (2017) Integrative modeling of gene and genome evolution roots the archaeal tree of life. *Proceedings of the National Academy of Sciences of the USA* 114, E4602–E4611.

Wilson, E.B. (1899) The structure of protoplasm. *Science* 10, 33–45.

Woese, C.R. (1968) The fundamental nature of the genetic code: Prebiotic interactions between polynucleotides and polyaminoacids and their derivatives. *Proceedings of the National Academy of Sciences of the USA* 59, 110–117.

Wright, P.E., Dyson, H.J. (1999) Intrinsically unstructured proteins: Re-assessing the protein structure-function paradigm. *Journal of Molecular Biology* 293, 321–331.

Wu, C.-Y., Travers, A. (2019) Modelling and DNA topology of compact 2-start and 1-start chromatin fibres. *Nucleic Acids Research* 47, 9902–9924.

Zamenhof, S., Shettles, L.B., Chargaff, E. (1950) Human deoxypentose nucleic acid. *Nature* 165, 756–757.

Zuo, Y., Steitz, T.A. (2015) Crystal structures of the *E. coli* transcription initiation complexes with a complete bubble. *Molecular Cell* 58, 534–540.

Tailpiece: Tree of life with one of Charles Jencks' DNA sculptures. Photograph by
Sarah Rodger, reproduced here with permission from Charles Jencks

Index